T0330999

Flexible Supply Chain

Communication is the glue that binds the diverse elements of the national economy. A flexible supply chain system creates postal communication agility and adaptability to respond to the changing technologies of the modern era. This book uses a systems-based approach of the tools and techniques of industrial engineering applied to a new concept of flexible supply chain systems, patterned after well-known and successful flexible manufacturing systems. It focuses on how industrial engineering can be used to achieve flexibility, resiliency, and efficiency in response to the needs of the global postal system.

Flexible Supply Chain: Industrial Engineering Optimization Modeling of the Postal Service System provides an understanding of the techniques of using flexibility to improve operations. It capitalizes on a framework of industrial engineering and focuses on a global chain of non-commodity products and services. By using flexibility as the foundation for efficiency, it ties in with the digital revolution of communication and retains an affiliation with community involvement. The book uses the DEJI Systems Model and the Triple C model as a structure of system productivity and creates a template through which other supply chains can be improved. The global supply chain is presently stressed and in need of new ideas and operational strategies.

This book is an ideal read for engineers working in manufacturing production, civil, mechanical, and other industries. It will be of interest to engineering managers and consultants as well as those involved with business management. University students and instructors will also find this book useful.

Systems Innovation Book Series

Series Editor: Adedeji B. Badiru

Systems Innovation refers to all aspects of developing and deploying new technology, methodology, techniques, and best practices in advancing industrial production and economic development. This entails such topics as product design and development, entrepreneurship, global trade, environmental consciousness, operations and logistics, introduction and management of technology, collaborative system design, and product commercialization. Industrial innovation suggests breaking away from the traditional approaches to industrial production. It encourages the marriage of systems science, management principles, and technology implementation. Particular focus will be the impact of modern technology on industrial development and industrialization approaches, particularly for developing economics. The series will also cover how emerging technologies and entrepreneurship are essential for economic development and society advancement.

Leadership Matters
An Industrial Engineering Framework for Developing and Sustaining Industry
Adedeji B. Badiru and Melinda Tourangeau

Systems Engineering
Influencing Our Planet and Reengineering Our Actions
Adedeji B. Badiru

Total Productive Maintenance, Second Edition
Strategies and Implementation Guide
Tina Agustiady and Elizabeth A. Cudney

Assessing Innovation
Metrics, Rubrics and Standards
Adedeji B. Badiru and Melinda Tourangeau

Integrating Artificial and Human Intelligence through Agent Oriented Systems Design
Michael E. Miller and Christina F. Rusnock

Handbook of Digital Innovation, Transformation, and Sustainable Development in a Post-Pandemic Era
Edited by M. Affan Badar, Ruchika Gupta, Priyank Srivastava, Imran Ali and Elizabeth A. Cudney

Flexible Supply Chain
Industrial Engineering Optimization Modeling of the Postal Service System
Adedeji B. Badiru

Flexible Supply Chain
Industrial Engineering Optimization
Modeling of the Postal Service System

Adedeji B. Badiru

CRC Press
Taylor & Francis Group
Boca Raton London New York

CRC Press is an imprint of the
Taylor & Francis Group, an **informa** business

Designed cover image: Shutterstock – Aun Photographer

First edition published 2025
by CRC Press
2385 NW Executive Center Drive, Suite 320, Boca Raton FL 33431

and by CRC Press
4 Park Square, Milton Park, Abingdon, Oxon, OX14 4RN

CRC Press is an imprint of Taylor & Francis Group, LLC

© 2025 Adedeji B. Badiru

ISBN: 978-1-032-61997-2 (hbk)
ISBN: 978-1-032-62068-8 (pbk)
ISBN: 978-1-032-62070-1 (ebk)

DOI: 10.1201/9781032620701

Typeset in Times
by Apex CoVantage, LLC

Dedicated to all postal workers who toil endlessly to ensure that postal products get to their intended destinations despite the modern digital and analog disruptions that often threaten the postal system.

Contents

Preface

This book uses a systems-based approach of the tools and techniques of industrial engineering applied to a new concept of flexible supply chain system, patterned after the well-known and successful flexible manufacturing systems. It focuses on how industrial engineering can be used to achieve flexibility, resiliency, and efficiency in response to the needs of the global postal system. The book illustrates the applicability of industrial engineering in all facets of endeavors including business, industry, government, the military, even academia. The specific application addressed in this book is the postal supply chain, which influences other commodity-based supply chains. Topics covered include classical tools and techniques of industrial engineering; the emergence of organized industry and linkages to commerce, business, and industry; elements of supply chain in general; specific needs of a postal supply chain; modeling of printed supply chain; efficiency, effectiveness, and productivity in the general communication industry; input-output relationships in the postal supply chain; systems optimization for commerce and industry; application of the Triple C model; application of the DEJI Systems Model; and the introduction of a new template for systems flexibility. There are pockets of inefficiency and ineffectiveness in any supply chain, due to a lack of systems accountability. Thus, the focus of this book is on how the diversity and versatility of industrial and systems engineering could be applied to affect process improvement onto a supply chain.

About the Author

Adedeji B. Badiru is Dean Emeritus, Graduate School of Engineering and Management at the US Air Force Institute of Technology (AFIT). While serving as Dean, he had oversight for planning, directing, and controlling operations related to granting doctoral and master's degrees for the US Air Force. He was previously Professor and Head of Systems Engineering and Management at AFIT, Professor and Department Head of Industrial Engineering at the University of Tennessee, and Professor of Industrial Engineering and Dean of the University College at the University of Oklahoma. He is a registered professional engineer (PE), a certified Project Management Professional (PMP), a Fellow of the Institute of Industrial & Systems Engineers (IISE), a Fellow of the Industrial Engineering and Operations Management Society (IEOM), a Fellow of Institute for Operations Research and Management Science (IORMS), and a Fellow of the Nigerian Academy of Engineering (Nigeria NAE). He is also a Program Evaluator for ABET. He holds a leadership certificate from the University Tennessee Leadership Institute. He has a BS in industrial engineering, an MS in mathematics, an MS in industrial engineering from Tennessee Technological University, and a PhD in industrial engineering from the University of Central Florida. His areas of interest include mathematical modeling, systems design and integration, project management and optimization, innovation management, and productivity improvement. He is a prolific author of over 45 books, over 40 book chapters, over 200 journal and magazine articles, and over 220 conference presentations. He is a member of several professional associations and scholastic honor societies. A world-renowned educator, he has won several awards for his teaching, research, administrative, and professional accomplishments. He is the 2020 recipient of the Lifetime Achievement Award from the Taylor & Francis publishing group. He was also a member of the AFIT team that won the 2019/2020 US Air Force Organizational Excellence Award. He is also the recipient of the 2022 BEYA career achievement award in the Government category and the 2024 Donald G. Newnan National Engineering Economy Teaching Excellence Award from the American Society for Engineering Education (ASEE). He is also received the 2024 Tau Beta Pi Distinguished Alumnus award. He holds a US trademark for the DEJI Systems Model for Design, Evaluation, Justification, and Integration. He is a member of the Fulbright Specialists Program.

Acknowledgments

Industrial engineering does it again. I acknowledge the support and contributions of my industrial engineering colleagues in our continuing efforts to promote the practice of industrial engineering globally. I thank Ms. Cindy Carelli and her very capable team in chaperoning the cause of industrial engineering in business, industry, academia, government, and the military. My special thanks go to my indomitable intellectual cheerleaders, Ms. Kathy Cretella and Ms. Melinda Tourangeau, for their incessant inquiries about when the manuscript was going to be finished. Their frequent prodding motivated me to keep on writing. This particular book required creative editorial finagling to get it right and in alignment with what the market needs. Thanks to all.

1 Systems View of the Supply Chain

INTRODUCTION

Whether subtle or explicit, supply has been a part of human existence from the beginning, not only for movement and delivery of goods, but also for our means of subsistence. Without some form of supply, life would be impossible. The following historical quote is aptly relevant in this regard:

> They turned their attention entirely to hunting and trapping. The forest *supplied* them abundantly with wild game for food.
> —John McClure, in "Sketches of a Western Adventure,"
> describing Simon Kenton exploring near the Ohio River in early 1788

Similarly, we recall the following eloquent reminder:

> To you I have given the deer, the bear, and all wild animals, and the fish that swim in the rivers, and the corn that grows in the fields.
> —Tenskwatawa, the Shawnee Prophet, 1826

As we can see, nature is the ultimate head of our global supply chain over the generations. The supply chain is all around us. It behooves us now to use our modern tools and techniques to maximize the outputs of our global supply chain, in all sectors of our economy and livelihood. Even supply and demand in the real estate sector are critical elements of our global supply chain, albeit not often recognized as such. Everything that we do in the global supply chain is done within the infrastructure of industrial, commercial, and residential construction. Thus, taking a systems view of the global supply chain is necessary for the sake or having a flexible, resilient, and adaptive system of moving goods and services from one point to another. A commonly accepted definition of a supply chain is provided here:

> The supply chain is the business function responsible for the connection and combination of activities related to the management, within and among organizations, of the flow of materials, products, information, commercial functions, and financial transactions.

It is not enough to proclaim a systems view of the world. We must actualize and practicalize the essence of a systems-centric world. At the end of the day, integration, in the context of the DEJI Systems Model presented in this book, is the critical key for actualizing a systems view. It is a proven fact that methodologies imported from one cultural system into another cultural system don't work well until they are

Integration of System Characteristics, Capabilities, Interests, Nuances, Limitations, etc.

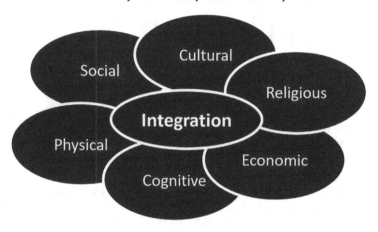

FIGURE 1.1   Integration of system characteristics across the spectrum of the supply chain.

adapted and integrated into the prevailing local practices of the target system. This can be observed around our present world, with vastly differing platforms of political systems, cultural systems, financial systems, educational systems, media systems, international trade systems, and other world-affective systems. In as much as the world is expected to work together, we must be cognizant of our cultural and operational differences. We must not be dismissive of where disconnections exist. Rather, we should confront, address, mitigate, or rectify the dissimilarities as we go through the structured tenets of the DEJI Systems Model of Design, Evaluation, Justification, and, finally, Integration. If this process is embraced and assimilated at the outset, from a systems perspective, we will have a better opportunity to make things (and people) work productively together in the long run. Figure 1.1 presents a pictorial representation of why integration (by any other name) is essential for achieving the ideals of a cohesive system. It is not expected that systems will align perfectly in all respects at all times. However, to whatever extent possible, we need to maximize the areas of overlap. This can be done through a combination of qualitative and quantitative techniques.

SUPPLY CHAIN OPERATIONS MANAGEMENT

Operations management is the core of any supply chain. Operations management, from a systems perspective, incorporates the elements, people, tools, and processes that make up the supply chain. For this purpose, both soft and hard sciences of management come into play. Specifically, systems integration is the foundation for operational success. Operational alignment is of particular interest to the Industrial Engineering and Operations Management (IEOM) society (https://ieomsociety.org). This commitment permeates all the principles of IEOM, as the organization works inclusively to align and integrate professional, operational, and educational interests around the world. The vision, mission, and values of IEOM are presented below to put this book in an operational context.

IEOM Vision: The IEOM Society International strives to be the premier global organization dedicated to the advancement of industrial engineering and operations management discipline for the betterment of humanity.

IEOM Mission: IEOM Society's core purpose is to globally foster critical thinking and its effective utilization in the field of Industrial Engineering (IE) and Operations Management (OM) by providing means to communicate and network among diversified people, especially in emerging countries, motivated by similar interests.

IEOM Values:

- Globalization, Diversity, and Inclusion
- Innovative and Entrepreneurial Thinking
- Student Empowerment and Mentoring
- Professional Development and Lifelong Learning
- Building Partnerships among Industry, Academia, and Government
- Sustainability and Community Development

A systems view of the world is required for people to live and work together for a common good around the world. *The profession of industrial engineering can help actualize a systems view of the world.* It is in this context and other operational strategies that the former Institute of Industrial Engineers (IIE) changed its name in 2016 to the Institute of Industrial and Systems Engineering (IISE). For decades, the institute was known as the American Institute of Industrial Engineers (AIIE). It was expanded in 1981 by dropping the "American" component of the name in recognition of the more international scope of the profession. The evolution of the profession over the years conveys the growing realization of a systems view of the world, beyond the traditional inward looking perspectives of organizations. Founded in 1948, IISE is the only international, nonprofit, professional society dedicated to advancing the technical and managerial excellence of industrial engineers. The profession of industrial engineering is recognized and applauded for advancing the practice of systems view globally. This is noted enough to embrace this author's cliché that "things operate better when industrial engineers are involved."

In the final analysis, everything works as a system. It is a systems world and everything around us is driven by a systems underpinning of how things work, both individually and together. Systems thinking facilitates the incorporation of all the elements and nuances in the operating environment. Nothing typifies this fact more than the profession of industrial and systems engineering (ISE). The foundational basis of industrial engineering is the appreciation of how "systems" permeate everything we do in business, industry, government, academia, the military, and other enterprises.

In response to operational disruptions of different modes, the world has learned to demonstrate, accept, embrace, and practice more flexibility in work, leisure, personal, and family activities. This calls for a more integrative systems view of everything we do. Under a systems perspective, whatever personal choices we make end up directly affecting other citizens. This realization of herd-to-herd systems interactions may help us nudge along those who are resistant, hesitant, or obstinate about community health initiatives.

It is the systems orientation that gives industrial engineering the flexibility, versatility, and professional mobility that it has. This is encapsulated in the very definition of the profession, as described in the section that follows. Our goal is to use our management tools to impact process improvement on the supply chain. In this regard, several factors of importance and relevance should be addressed:

- Supply disruption risk;
- Complexity of the supply chain;
- Supplier proficiency;
- Supply sourcing cost;
- Technology availability and maturity;
- System affordability;
- Market uncertainty;
- Customer needs and wants;
- Demand variation;
- Inventory depletion rate;
- Product attributes and adaptability;
- Innovation anchor and management;
- Workforce availability and reliability;
- Process agility;
- Transportation infrastructure;
- Supply communication, cooperation, and coordination;
- Production cycle and delivery modes;
- Product design and malleability.

It is almost an inexhaustible list. Only the versatile tools and techniques of industrial engineering can handle the diversity of operational needs.

INDUSTRIAL ENGINEERING IN THE SUPPLY CHAIN

Industrial engineering facilitates innovation in the supply chain. A common question is "who is an industrial engineer?" Because the profession of industrial engineering is very diverse, flexible, and widely encompassing, it is often challenging to define the wide scope of industrial engineering in a few sentences:

> An industrial engineer is someone who is concerned with the design, installation, and improvement of integrated systems of people, materials, information, equipment, and energy by drawing upon specialized knowledge and skills in the mathematical, physical, and social sciences, together with the principles and methods of engineering

analysis and design to specify, predict, and evaluate the results to be obtained from such systems.

Systems implementation drives innovation. Thus, by inference, industrial engineering principles are foundational for driving, anchoring, and sustaining innovation, from a systems platform. Whatever your industry or business, you are operating in a system. The comprehensive definition of industrial engineering says it aptly. In the perspective of industrial engineering, everything is a system. The more things are viewed as a system, the better we can have a consistent and comprehensive handle on its operations. Industrial and systems engineers (ISEs) are perhaps the most preferred engineering professionals because of their ability to manage complex organizations (Oke, 2014). They are trained to design, develop, and install optimal methods for coordinating people, materials, equipment, energy, and information. The integration of these resources is needed in order to create products and services in a business world that is becoming increasingly complex and globalized. Industrial and systems engineers oversee management goals and operational performance. Their aims are the effective management of people, coordinating techniques in business organizations, and adapting technological innovations toward achieving increased performance. They also stimulate awareness of the legal, environmental, and socioeconomic factors that have a significant impact on engineering systems. Industrial and systems engineers can apply creative values in solving complex and unstructured problems in order to synthesize and design potential solutions and organize, coordinate, lead, facilitate, and participate in teamwork. They possess good mathematical skills, a strong desire for organizational performance, and a sustained drive for organizational improvement.

In deriving efficient solutions to manufacturing, organizational, and associated problems, ISEs analyze products and their requirements. They utilize mathematical techniques such as operations research (OR) to meet those requirements, and to plan production and information systems. They implement activities to achieve product quality, reliability, and safety by developing effective management control systems to meet financial and production planning needs. Systems design and development for the continual distribution of the product or service is also carried out by ISEs to enhance an organization's ability to satisfy their customers. Industrial and systems engineers focus on optimal integration of raw materials available, transportation options, and costs in deciding plant location. They coordinate various activities and devices on the assembly lines through simulations and other applications.

The organization's wage and salary administration systems and job evaluation programs can also be developed by them, leading to their eventual absorption into management positions. They share similar goals with health and safety engineers in promoting product safety and health in the whole production process through the application of knowledge of industrial processes and such areas as mechanical, chemical, and psychological principles. They are well grounded in the application of health and safety regulations while anticipating, recognizing, and evaluating hazardous conditions and developing hazard-control techniques.

Industrial and systems engineers can assist in developing efficient and profitable business practice by improving customer services and the quality of products. This would improve the competitiveness and resource utilization in organizations. From another perspective, ISEs are engaged in setting traditional labor or time standards and in the redesign of organizational structure in order to eliminate or reduce some forms of frustration or waste in manufacturing. This is essential for the long-term survivability and the health of the business.

Another aspect of the business that the ISEs could be useful in is making work safer, easier, more rewarding, and faster through better designs that reduce production cost and allow the introduction of new technologies. This improves the lifestyle of the populace by making it possible for them to afford and use technological advanced goods and services. In addition, they offer ways of improving the working environment, thereby improving efficiencies and increasing cycle time and throughput, and helping manufacturing organizations to obtain their products more quickly. Also, ISEs have provided methods by which businesses can analyze their processes and try to make improvements upon them. They focus on optimization – doing more with less – and help to reduce waste in the society. ISEs give assistance in guiding the society and business to care more for their workforce while improving the bottom line.

Perhaps the first classic and widely accepted definition of industrial engineering (IE) was offered by the then American Institute of Industrial Engineering (AIIE) in 1948. Others have extended the definition. Industrial engineering is uniquely concerned with the analysis, design, installation, control, evaluation, and improvement of sociotechnical systems in a manner that protects the integrity and health of human, social, and natural ecologies. A sociotechnical system can be viewed as any organization in which people, materials, information, equipment, procedures, and energy interact in an integrated fashion throughout the life cycles of its associated products, services, or programs. Through a global system's perspective of such organizations, industrial engineering draws upon specialized knowledge and skills in the mathematical, physical, and social sciences, together with the principles and methods of engineering analysis and design, to specify the product and evaluate the results obtained from such systems, thereby assuring such objectives as performance, reliability, maintainability, schedule adherence, and cost control.

There are five general areas of industrial and systems engineering. Each of these areas specifically makes out some positive contributions to the growth of industrial and systems engineering. The first area comprises sociology and economics. The combination of the knowledge from these two areas helps in the area of supply chain. The second area is mathematics, which is a powerful tool of ISEs. Operations research is an important part of this area. The third area is psychology, which is a strong pillar for ergonomics. Accounting and economics both constitute the fourth area. These are useful subjects in the area of engineering economics. The fifth area is computer. Computers are helpful in CAD/CAM, which is an important area of industrial and systems engineering.

According to the International Council on Systems Engineering (INCOSE), systems engineering is an interdisciplinary approach and means to enable the realization of successful systems. Such systems can be diverse, encompassing people and organizations, software and data, equipment and hardware, facilities and materials, and services and techniques. The system's components are interrelated and employ organized interaction toward a common purpose. From the viewpoint of INCOSE, systems engineering focuses on defining customer needs and required functionality early in the development cycle, documenting requirements, and then proceeding with design synthesis and systems validation while considering the complete problem. The philosophy of systems engineering teaches that attention should be focused on what the entities do before determining what the entities are. A good example to illustrate this point may be drawn from the transportation system. In solving a problem in this area, instead of beginning the problem-solving process by thinking of a bridge and how it will be designed, the systems engineer is trained to conceptualize the need to cross a body of water with certain cargo in a certain way.

The systems engineer then looks at bridge design from the point of view of the type of bridge to be built. For example, is it going to have a suspension or superstructure design? From this stage they would work down to the design detail level where the systems engineer gets involved, considering foundation soil mechanics and the placement of structures. The contemporary business is characterized by several challenges. This requires the ISEs to have skills, knowledge, and technical know-how in the collection, analysis, and interpretation of data relevant to problems that arise in the workplace. This places the organization well above the competition.

The radical growth in global competition, constantly and rapidly evolving corporate needs, and the dynamic changes in technology are some of the important forces shaping the world of business. Thus, stakeholders in the economy are expected to operate within a complex but ever-changing business environment. Against this backdrop, the dire need for professionals who are reliable, current, and relevant becomes obvious. Industrial and systems engineers are certainly needed in the economy for bringing about radical change, value creation, and significant improvement in productive activities.

The ISE must be focused and have the ability to think broadly in order to make a unique contribution to the society. To complement this effort, the organization itself must be able to develop effective marketing strategies (aided by a powerful tool, the Internet) as a competitive advantage so that the organization could position itself as the best in the industry.

The challenges facing the ISE may be divided into two categories: Those faced by ISEs in developing and underdeveloped countries, and those faced by engineers in the developed countries. In the developed countries, there is a high level of technological sophistication that promotes and enhances the professional skills of the ISE. Unfortunately, the reverse is the case in some developing and underdeveloped countries. Engineers in underdeveloped countries, for instance, rarely practice technological development, possibly owing to the high level of poverty in such environments. Another reason that could be advanced for this is the shortage of skilled manpower in the engineering profession that could champion technological breakthrough similar to the channels operated by the world economic powers. In addition, the technological development of nations could be enhanced by the formulation of active research teams. Such teams should be focused with the aims of solving practical industrial problems. Certain governments in advanced countries encourage engineers (including ISEs) to actively participate in international projects funded by government or international agencies. For the developing and underdeveloped countries, this benefit may not be gained by the ISE until the government is challenged to do so in order to improve on the technological development of the country.

Challenges before a community may be viewed from the perspective of the problem faced by the inhabitants of that community. As such, they could be local or global. Local challenges refer to the need must be satisfied by the engineers in that community. These needs may not be relevant to other communities, for example, the ISE may be in a position to advise the local government chairman of a community on the disbursement of funds on roads within the powers of the local government. Decision-science models could be used to prioritize certain criteria, such as the number of users, the economic indexes of the various towns and villages, the level of business activities, the number of active industries, the length of the road, and the topography or the shape of the road.

Soon after graduation, an ISE is expected to tackle a myriad of social, political, and economic problems. This presents a great challenge to the professionals who live in a society where these problems exist. Consider the social problems of electricity generation, water provision, flood control, and so on. The ISE in a society where these problems exist is expected to work together with other engineers in order to solve these problems. They are expected to design, improve on existing designs, and install integrated systems of workers, materials, and equipment so as to optimize the use of resources. For electricity distribution, the ISE should be able to develop scientific tools for the distribution of power generation as well as for the proper scheduling of the maintenance tasks to which the facilities must be subjected.

The distribution network should minimize the cost. Loss prevention should be a key factor to consider. As such, the quality of the materials purchased for maintenance should be controlled, and a minimum acceptable standard should be established. In solving water problems, for instance, the primary distribution route should be a major concern. The ISE may need to develop reliability models that could be applied to predict the life of components used in the system. The scope of activities of the ISE should be wide enough for them to work with other scientists in the health sector on modeling and control of diseases caused by water-distribution problems. The ISE should be able to solve problems under uncertain conditions and limited budgets.

The ISE can work in a wide range of industries, such as the manufacturing, logistics, service, and defense industries. In manufacturing, the ISE must ensure that the equipment, manpower, and other resources in the process are integrated in such a manner that efficient operation is maintained and continuous improvement is ensured. The ISE functions in the logistics industry through the management of supply chain systems (e.g., manufacturing facilities, transportation carriers, distribution hubs, retailers) to fulfill customer orders in the most cost-effective way. In the service industry, the ISE provides consultancies in areas related to organizational effectiveness, service quality, information systems, project management, banking, service strategy, and so on. In the defense industry, the ISE provides tools to support the management of military assets and military operations in an effective and efficient manner. The ISE works with a variety of job titles. The typical job titles of an ISE graduate include industrial engineer, manufacturing engineer, logistics engineer, supply chain engineer, quality engineer, systems engineer, operations analyst, management engineer, and management consultant.

Experiences in the United States and other countries show that a large proportion of ISE graduates work in consultancy firms or as independent consultants, helping companies to engineer processes and systems to improve productivity, effect efficient operation of complex systems, and manage and optimize these processes and systems.

After completing their university education, ISEs acquire skills from practical exposure in an industry. Depending on the organization that an industrial or systems engineer works for, the experience may differ in depth or coverage. The trend of professional development in industrial and systems engineering is rapidly changing in recent times. This is enhanced by the ever-increasing development in the information, communication and technology (ICT) sector of the economy.

Industrial and systems engineering is methodology-based and is one of the fastest growing areas of engineering. It provides a framework that can be focused on any area of interest, and incorporates inputs from a variety of disciplines, while maintaining the engineer's familiarity and grasp of physical processes. The honor of discovering industrial engineering belongs to a large number of individuals. The eminent scholars in industrial engineering are Henry Gantt (the inventor of the Gantt chart) and Lillian Gilbreth (a coinventor of time and motion studies). Some other scientists have also contributed immensely to its growth over the years. The original application of industrial engineering at the turn of the century was in manufacturing a technology-based orientation, which gradually changed with the development of OR, cybernetics, modern control theory, and computing power.

Computers and information systems have changed the way industrial engineers do business. The unique competencies of an ISE can be enhanced by the powers of the computer. Today, the fields of application have widened dramatically, ranging from the traditional areas of production engineering, facilities planning, and material handling to the design and optimization of more broadly defined systems. An ISE is a versatile professional who uses scientific tools in problem solving through a holistic and integrated approach. The main objective of an ISE is to optimize performance through the design, improvement, and installation of integrated of human, machine, and equipment systems. The uniqueness of industrial and systems engineering among engineering disciplines lies in the fact that it is not restricted to technological or industrial problems alone. It also covers nontechnological or non-industry-oriented problems. The training of ISEs positions them to look at the total picture of what makes a system work best. They question themselves about the right combination of human and natural resources, technology and equipment, and information and finance. The ISEs make the system function well. They design and implement innovative processes and systems that improve quality and productivity, eliminate waste in organizations, and help them to save money or increase profitability.

Industrial and systems engineers are the bridges between management and engineering in situations where scientific methods are used heavily in making managerial decisions. The industrial and systems engineering field provides the theoretical and intellectual framework for translating designs into economic products and services, rather than the fundamental mechanics of design. Industrial and systems engineering is vital in solving today's critical and complex problems in manufacturing, distribution of goods and services, health care, utilities, transportation, entertainment, and the environment. The ISEs design and refine processes and systems to improve quality, safety, and productivity. The field provides a perfect blend of technical skills and people orientation. An industrial engineer addresses the overall system performance and productivity, responsiveness to customer needs, and the quality of the products or services produced by an enterprise. Also, they are the specialists who ensure that people can safely perform their required tasks in the workplace environment. Basically, the field deals with analyzing complex systems, formulating abstract models of these systems, and solving them with the intention of improving system performance.

The discussions under this section mainly consist of some explanations of the areas that exist for industrial and systems engineering programs in major higher institutions the world over.

HUMAN SYSTEMS IN THE SUPPLY CHAIN

Fredrick Taylor's 1919 landmark treatise, *The Principles of Scientific Management*, discussed key methods of improving human productivity in systems (Miller et al., 2014). These techniques included selecting individuals compatible with their assigned task, tasking an appropriate number of individuals to meet time demands, providing training to the individuals, and designing work methods and implements with productivity in mind (Taylor, 1919). Taylor asserted that through improvements, which resulted in increased human productivity, the cost of manpower could be significantly reduced to improve "national efficiency." That is, the ratio of national manufacturing output to the cost of manufacture could be significantly increased. Over the intervening decades between Taylor's *Principles* and today, the domains of *manpower, personnel, training*, and *human factors* have evolved to independently address each of these respective needs. For instance, the manpower domain defines appropriate staffing levels, while the personnel domain addresses the recruitment, selection, and retention of individuals to achieve those staffing levels. The training domain insures that each worker has the appropriate knowledge, skills, and abilities to perform their assigned tasks. And the human factors domain focuses on the design and selection of tools that effectively augment human capabilities to improve the productivity of each worker.

Taylor demonstrated, using concrete examples taken from industry (e.g., a steel mill), that selecting appropriate individuals for a task (e.g., selecting physically strong individuals for moving heavy pig iron) could allow more work to be done by fewer people. Training the individuals with the best work practices (e.g., training and incentivizing individuals to avoid nonproductive steps) also increased individual worker productivity. Likewise, improving tool selection (e.g., using larger shovels for moving light coke and smaller shovels for moving heavier coal) had a similar effect. While each of these observations were readily made in early 20th-century steel mills, an increasing amount of today's work is cognitive rather than physical in nature – and it is often performed by distributed networks of individuals working in cyberspace rather than by workers physically collocated on a factory floor. The effect of this trend is to reduce the saliency of Taylor's examples that largely focused on manual tasks (Miller et al., 2014).

It has been known for over a century that the domains of personnel, training, and human factors affect productivity and required manpower. Yet the term human systems integration (HSI) is relatively new, so what is HSI?

The concept of HSI emerged in the early 1980s, starting in the US Army, as a more modern construct for holistically considering the domains of manpower, personnel, training, and human factors, among others. The concept has gained emphasis, both within military acquisition and the systems engineering community. The HSI concept is based on the axiom that a human-centered focus throughout the design and operation of systems will ensure the following:

- Effective human-technology interfaces are incorporated in systems.
- Required levels of sustained human performance are achieved.
- Demands upon personnel resources, skills, and training are economical.
- Total system ownership costs are minimized.
- The risk of loss or injury to personnel, equipment, and/or the environment is minimized.

HSI deals with the complexity inherent in the problem space of human performance in systems by decomposing human-related considerations into focus areas or domains (i.e., HSI analysis), which essentially form a checklist of issues that need to be considered. These domains are often aligned with specific scientific disciplines or functional areas within organizations and may vary based on the perspective and needs of individual system developers and/or owners. Equally important, the HSI concept assumes the following corollary: Domains are interrelated and must be "rolled-up" and viewed holistically (i.e., HSI synthesis) to effectively understand and evaluate anticipated human performance in systems. What emerges is a view of HSI as a recursive cycle of analysis, synthesis, and evaluation, yielding HSI domain solution sets.

The decomposition of human-related considerations into domains is necessarily human-made and largely a matter of organizational convenience. Accordingly, we will not argue for the existence of an exhaustive and mutually exclusive set of HSI domains. Instead, what follows is a set of domains and their respective descriptions that have proven intuitive to an international audience:

- *Manpower/personnel* concerns the number and types of personnel required and available to operate and maintain the system under consideration. It considers aptitudes, experience, and other human characteristics, including body size, strength, and less tangible attributes necessary to achieve optimum system performance. This domain also includes the necessary selection processes required for matching qualified personnel to the appropriate task, as well as tools to assess the number of individuals necessary to achieve a desired level of system performance.
- *Training* embraces the specification and evaluation of the optimum combination of instructional systems; education; and on-the-job training required to develop the knowledge, skills, and abilities (e.g., social/team building abilities, soft skills, competencies) needed by the available personnel to

operate and maintain the system under consideration to a specified level of effectiveness under the full range of operating considerations.

- *Human factors* are the cognitive, physical, sensory, and team dynamic abilities required to perform system-specific operational, maintenance, and support job tasks. This domain covers the comprehensive integration of human characteristics into system design, including all aspects of workstation and workspace design and system safety. The objective of this domain is to maximize user efficiency while minimizing the risk of injury to personnel and others.

- The *safety and health* domain includes applying human factors and engineering expertise to minimize safety risks occurring as a result of the system under consideration being operated or functioning in a normal or abnormal manner. System design features should serve to minimize the risk of injury, acute or chronic illness, and/or discomfort of personnel who operate, maintain, or support the system. Likewise, design features should mitigate the risk for errors and accidents resulting from degraded job performance. Prevalent safety and health issues include noise, chemical safety, atmospheric hazards (including those associated with confined space entry and oxygen deficiency), vibration, ionizing and non-ionizing radiation, and human factors issues that can create chronic disease and discomfort such as repetitive motion injuries. Human factors stresses that create risk of chronic disease and discomfort overlap with occupational health considerations. These issues directly impact crew morale.

- The *organizational and social* domain applies tools and techniques drawn from relevant information and behavioral science disciplines to design organizational structures and boundaries around clear organizational goals to enable people to adapt an open culture, improving sharing and trust between colleagues and coalition partners. This domain focuses on reducing the complexity of organizations. Although pertinent to all organizations, this domain is particularly germane to modern systems employing network-enabled capabilities as successful operation of these systems requires trust and confidence to be built between people in separate organizations and spatial locations who need to collaborate on a temporary basis without the opportunity to build personal relationships.

- *Other domains* have been others proposed and are worth brief mention. While the above areas have consistently been included in the HSI literature, other domains could include *personnel survivability, habitability*, and the *environment*. Personnel survivability, a military focus area, assesses designs that reduce risk of fratricide, detection, and the probability of being attacked and enable the crew to withstand human-made or natural hostile environments without aborting the mission or suffering acute or chronic illness or disability/death. Habitability addresses factors of living and working conditions that are necessary to sustain the morale, safety, health, and comfort of the user population which contribute directly to personnel effectiveness. Finally, environmental design factors concern water, air and land pollution and their interrelationships with system manufacturing, operation and disposal.

Human factors engineering is a practical discipline dealing with the design and improvement of productivity and safety in the workplace. It concerns the relationship of manufacturing and service technologies interacting with humans. Its focus is not restricted to manufacturing alone – it extends to service systems as well. The main methodology of ergonomics involves the mutual adaptation of the components of human-machine-environment systems by means of human-centered design of machines in production systems. Ergonomics studies human perceptions, motions, workstations, machines, products, and work environments.

Today's ever-increasing concerns about humans in the technological domain make this field very appropriate. People in their everyday lives or in carrying out their work activities create many of the human-made products and environments for use. In many instances, the nature of these products and environments directly influences the extent to which they serve their intended human use. The discipline of human factors deals with the problems and processes that are involved in man's efforts to design these products and environments so that they optimally serve their intended use by humans. This general area of human endeavor (and its various facets) has come to be known as human factors engineering or, simply, human factors, biomechanics, engineering psychology, or ergonomics.

Operations research specifically provides the mathematical tools required by ISEs in order to carry out their task efficiently. Its aims are to optimize system performance and predict system behavior using rational decision making, and to analyze and evaluate complex conditions and systems.

This area of industrial and systems engineering deals with the application of scientific methods in decision making, especially in the allocation of scarce human resources, money, materials, equipment, or facilities. It covers such areas as mathematical and computer modeling and information technology. It could be applied to managerial decision making in the areas of staff and machine scheduling, vehicle routing, warehouse location, product distribution, quality control, traffic light phasing, and police patrolling. Preventive maintenance scheduling, economic forecasting, experiment design, power plant fuel allocation, stock portfolio optimization, cost-effective environmental protection, inventory control, and university course scheduling are some of the other problems that could be addressed by employing OR.

Subjects such as mathematics and computer modeling can forecast the implications of various choices and identify the best alternatives. The OR methodology is applied to a wide range of problems in both public and private sectors. These problems often involve designing systems to operate in the most effective way. Operations research is interdisciplinary and draws heavily on mathematics. It exposes graduates in the field of industrial and systems engineering to a wide variety of opportunities in areas such as pharmaceuticals, ICT, financial consultancy services, manufacturing, research, logistics and supply chain management, and health. These graduates are employed as technical analysts with prospects for managerial positions. Operations research adopts courses from computer science, engineering management, and other engineering programs to train students to become highly skilled in quantitative and qualitative modeling and the analysis of a wide range of systems-level decision problems. It focuses on productivity, efficiency, and quality. It also affects the creative utilization of analytical and computational skills in problem solving, while increasing the knowledge necessary to become truly competent in today's highly competitive business environment. Operations research has had a tremendous impact on almost every facet of modern life, including marketing, the oil and gas industry, the judiciary, defense, computer operations, inventory planning, the airline system, and international banking. It is a subject of beauty whose applications seem endless.

The aim of studying artificial intelligence (AI), as a system, is to understand how the human mind works, thereby fostering leading to an appreciation of the nature of intelligence, and to engineer systems that exhibit intelligence. Some of the basic keys to understanding intelligence are vision, robotics, and language. Other aspects related to AI include reasoning, knowledge representation, natural language generation (NLG), genetic algorithms, and expert systems. Studies on reasoning have evolved from the following dimensions: Case-based, nonmonotonic, model, qualitative, automated, spatial, temporal, and common sense. For knowledge representation, knowledge bases are used to model application domains and to facilitate access to stored information. Knowledge representation originally concentrated around protocols that were typically tuned to deal with relatively small knowledge bases but that provided powerful and highly expressive reasoning services. Natural language generation systems are computer software systems that produce texts in English and other human languages, often from nonlinguistic input data. Natural language generation systems, like most AI systems, need substantial amounts of knowledge that is difficult to acquire. In general terms, these problems were due to the complexity, novelty, and poorly understood nature of the tasks our systems attempted, and were worsened by the fact that people write so differently. A genetic algorithm is a search algorithm based on the mechanics of natural selection and natural genetics. It is an iterative procedure that maintains a population of structures that are candidate solutions to specific domain challenges. During each generation the structures in the current population are rated for their effectiveness as solutions, and on the basis of these evaluations, a new population of candidate structures is formed by using specific genetic operators such as reproduction, crossover, and mutation. An expert system is a computer software that can solve a narrowly defined set of problems using information and reasoning techniques normally associated with a human expert. It could also be viewed as a computer system that performs at or near the level of a human expert in a particular field of endeavor.

A model is a simplified representation of a real system or phenomenon. Models are abstractions revealing only the features that are relevant to the real system behavior under study. In industrial and systems engineering, virtually all areas of the disciplines have concepts that can be modeled in one form or the other. In particular, mathematical models are elements, concepts, and attributes of a real system represented by using mathematical symbols. Models are powerful tools for predicting the behavior of a real system by changing some items in the models to detect the reaction of changes in the behavior of other variations. They provide frames of reference by which the performance of the real system can be measured. They articulate abstractions, thereby enabling us to distinguish between relevant and irrelevant features of the real system. Models are prone to manipulations more easily in a way that the real systems are often not.

In order to survive in the competitive environment, significant changes should be made in the ways of preparing organizations' design, and manufacturing, selling, and servicing their goods and commodities. Manufacturers are committed to continuous improvement in product design, defect levels, and costs. This is achieved by fusing designing, manufacturing, and marketing into a complete whole. A manufacturing system consists of two parts: Its science and automation. Manufacturing science refers to investigations on the processes involved in the transformation of raw materials into finished products. This involves the traditional aspects. Traditionally, manufacturing science may refer to the techniques of work-study, inventory systems, material-requirement planning, and so on. On the other hand, the automation aspect of manufacturing covers issues like e-manufacturing, the Toyota system, the use of computer-assisted manufacturing systems (NC, CNC, and DNC), automated material handling systems, group technology, flexible manufacturing systems, process planning and control, and so on.

Industrial and systems engineering students conduct research in the areas of manufacturing in combination with courses in finance, manufacturing processes, and personnel management. They also do research in manufacturing-design projects. This exposes the students to a manufacturing environment with activities in the design or improvement of manufacturing systems, product design, and quality.

Recent years have experienced increasing use of statistics in the industrial and systems engineering field. Industrial and systems engineers need to understand the basic statistical tools to function in a world that is becoming increasingly dependent on quantitative information. This clearly shows that the interpretation of practical and research results in industrial and systems engineering depends to a large extent on statistical methods. Statistics is used in almost every area relevant to these fields. It is utilized as a tool for evaluating economic data in "financial engineering." For this reason, ISEs are exposed to statistical reasoning early in their careers. Industrial and systems engineers also employ statistical techniques to establish quality control techniques. This involves detecting an abnormal increase in defects, which reflects equipment malfunction. The question of what, how, and when do we apply statistical techniques in practical situations and how to interpret the results are answered in the topics related to statistics.

The impact of industrial and systems of computers on engineering is complex and many-sided. The practitioners of data analysis in industrial and systems engineering rely on a computer as it is an important and powerful tool for collecting, recording, retrieving, analyzing simple and complex problems, as well as distributing huge information in industrial and systems engineering. It saves countless years of tedious work by the ISEs. The computer removes the necessity to monitor and control tedious and repetitive processes. Despite the importance of computers, its potential is so little explored that its full impact is yet to be realized. There are several powerful computer programs that can reduce the complexity of solving engineering problems.

PRODUCTIVITY, EFFICIENCY, AND EFFECTIVENESS

According to Miller et al. (2014), Taylor's original focus was on "national efficiency," but to what end(s) should HSI be oriented? System designers and impending system owners, by necessity, must compare potential solutions to exploit opportunities and select the best solution. What is the yardstick for identifying the "best" HSI domain solution set?

To answer this question, we focus on three highly interrelated but distinct terms: Productivity, efficiency, and effectiveness. *Productivity* is the rate at which goods or services are produced and is typically specified as the number of completed elements per unit of labor. *Efficiency* is the ratio of useful output to the total input in any system. Therefore, productivity might be defined as the efficiency of the human operator. *Effectiveness* refers to the ability to produce a desired effect.

We now provide an illustration of the differences among these terms. Suppose that a manufacturing system is defined to be effective if it permits a company to produce an article at a cost that allows the article to be sold at a profit. If an operator is solely responsible for production of this article and is paid based on a fixed wage, this manufacturing system will require a minimum level of productivity from the operator to be effective. Increasing the operators' productivity permits the cost of the operator to be distributed across a larger number of articles. Consequently, the cost per article associated with the operator is reduced. Restated, productivity or human efficiency can contribute to effectiveness of the manufacturing system. In traditional human-centered disciplines, such as human factors engineering, it is common to measure efficiencies such as the time required to complete a task to understand or quantify the quality of an interface.

However, productivity and efficiency are not synonymous with effectiveness. Many other factors – including factors that are both internal and external to the system – can influence the effectiveness of the system. Returning to our example, if the raw materials or market demand for the article produced by the operator is not present, the manufacturing system will not be effective regardless of the efficiency of the operator. These external influences thus render the operators' efficiency meaningless with regard to overall system effectiveness. Further, if the equipment required to process the parts outside the operators' workstation cannot match the operators' production rate, the operator's efficiency again may not have an effect upon the overall system effectiveness. Therefore, effectiveness might be influenced, but is not controlled, by productivity or operator efficiency.

Effectiveness is therefore a more valuable measure than productivity and efficiency because it assesses the degree to which a system serves its organizational purpose as well as harmonizes with other systems and its environment. Given HSI's focus on maximizing effectiveness at minimum cost – that is, maximizing the ratio of effectiveness to cost – *cost effectiveness* appears to be a naturally intuitive, unidimensional measure of merit for comparing individual HSI domain solution sets. Since money is the closest thing we currently have to a universal means of exchange, it is possible to assign costs to each HSI domain (Miller et al., 2014). Thus, cost effectiveness allows comparisons of the performance of HSI domain solution sets in terms of the desired system emergent properties while also capturing the input parameter of the summation of the respective domain-related costs.

While cost effectiveness is a useful measure, it may not always result in universally better systems in terms of the owning organization; for example, depicting a SOI and three sibling systems, each consisting of a personnel subsystem and a technological subsystem. The SOI and sibling systems, in turn, are components of a larger containing system (i.e., the parent organization). If we solely focus on the SOI, we would seek the joint optimization of the personnel and technological subsystems within the SOI, thereby maximizing its cost effectiveness and contributing positively to the containing system's objectives. Now, let us assume that the SOI and its sibling systems must share a common personnel resource pool. It is possible in maximizing the cost-effectiveness of the SOI to have unintended downstream effects on the personnel subsystems of the sibling systems. These downstream effects could result in decreased effectiveness of the sibling systems. In aggregate, maximizing the cost effectiveness of the SOI may actually result in a net negative contribution towards achieving the containing system's objective! Such a scenario illustrates the need to also consider the "net contribution" of a HSI domain solution set.

Timing is everything, or so the saying goes. The greatest value is obtained by the early involvement of HSI specialists. Between 70% and 90% of life cycle costs are already locked in by the end of conceptual system design. Given that 80% of life cycle costs are HSI related and 40%–60% is attributable to the manpower, personnel, and training domains alone, the longer one waits to begin addressing HSI, the more negative impact will be shown on total life cycle costs. Additionally, since human performance contributes significantly to system effectiveness, the only question is whether HSI will be paid for most affordably in advance, or at much greater expense after a mature system design reveals significant problems. The earlier an HSI investment can be made, the greater will be its return. However, it is important to remember that there will be benefits of incorporating HSI at any point in the design maturity, as long as it precedes the final design.

All the aforementioned characteristics of the world of systems point to the need for guiding models for understanding, managing, and advancing integrated systems of people, processes, and tools. This is one of the basic foundations for the structural framework of the DEJI Systems Model. In many human endeavors, things are often naturally "unstructured." The DEJI Systems Model can impart structure in many systems, whether human or digital.

WORKFORCE WELFARE

The basic element of a systems approach is to recognize people in the supply chain system. No matter how much a system is automated, people are still required to oil the operational gears of the system. If we take care of the people, the people will take care of the system and the supply chain will be well honed. In this regard, having a commitment to workforce welfare is required for a successful supply chain. Administrators, managers, and supervisors need to recognize and appreciate the wear and tear on the workforce as a result of the stress of the job. Badiru and Barlow (2020) present a call for new innovations in workforce development and redevelopment, following operational disruptions, such as the one caused by COVID-19. The workforce disruption caused by COVID-19 has necessitated a new focus on the challenges of workforce development in the State of Ohio. The decimation of productivity caused by COVID-19 requires not only the traditional strategies of workforce development, but also the uncharted territory of workforce redevelopment and preservation.

Reporting during COVID-19 indicates a precipitous decline in the ability of the workforce to continue to contribute to economic development and vitality of the state during the lockdown. When businesses open again, it will be necessary for workers to relearn their jobs to return to the level of proficiency and efficiency needed to move the state's economy forward. The technical topic of learning curve analysis postulates that performance improves with repeated cycles of operations. Whenever work performance is interrupted for a prolonged period, as we are currently experiencing, the processes of natural forgetting or technical regressing set in. To offset this decline, direct concerted efforts must be made beyond anything we have experienced before. This urgency to recover the economy led to our call for new innovations in workforce development and redevelopment. We cannot be lackadaisical in leaving things to the normal process of regaining form, routine, and function.

Typically, we erroneously focus on technical tools as the embodiment of innovation. But more often than not, process innovations might be just as vital. Workforce development, in particular, is more process development than tool development. There are numerous human factors strategies that can enhance the outcomes of workforce development. Some of the innovations we recommend in this regard include paying attention to the hierarchy of needs of the worker (primarily safety in our current world), recognizing the benefits of diversity, elevating the visibility of equity, instituting efforts to negate adverse aspects of cultural bias, and appreciating the dichotomy of socio-economic infrastructure. While not too expensive to implement, these innovative strategies can be tremendously effective.

Workforce redevelopment is a topic not very often discussed, but COVID-19 brings its importance to the forefront. Redevelopment will be needed not only to boost the quantity of the productive capacity, but also to restore and augment the capability, availability, and reliability of the workforce beyond the previous performance yardstick.

The greatest challenge in a COVID-19 environment is workforce preservation. We don't think there will be a post-COVID-19 environment in the near future. Coronavirus, the virus that causes COVID-19, is something we may have to contend with cyclically into the foreseeable future. How do we preserve the workforce in such a persistent COVID environment? Preservation of a well-developed workforce can only be assured through innovative health and safety safeguards, as well as new organizational processes and procedures that will take a thorough understanding of the recurring risks that may be posed by virus outbreaks. A workforce member who becomes ill or decides to leave an organization is a workforce member that we fail to preserve. Typically, a society addresses safety and security as necessary social mandates. Our postulation here is that we need to elevate that perception to the level of workforce necessity. A workforce that is well educated and well developed but stymied by the implications of a virus cannot be a productive workforce that contributes to the continued economic development of a region or state. Institutions of higher learning should continue to partner in addressing new innovations in workforce development, redevelopment, welfare, and preservation. More widespread partnerships are needed in this effort that bodes well for the economic health and vitality of any region in the overall supply chain.

REFERENCES

Badiru, A., & Barlow, C. (2020, May 15). Developing workforce in era of COVID-19. *Dayton Daily News Newspaper*, B7.

Miller, M. E., Colombi, J., & Tvaryanas, A. (2014). Human systems integration. In A. B. Badiru (Ed.), *Handbook of industrial and systems engineering* (2nd ed., pp. 197–216). CRC Press.

Oke, S. A. (2014). An overview of industrial and systems engineering. In A. B. Badiru (Ed.), *Handbook of industrial and systems engineering* (2nd ed., pp. 185–196). CRC Press.

Taylor, F. W. (1919). *The principles of scientific management*. Harper and Brothers Publishers.

2 Process Improvement in the Supply Chain

INTRODUCTION

Process improvement is the ultimate goal in the global supply chain. Because of the diverse, flexible, and comprehensive systems-based methodologies of industrial engineering, the discipline offers the best views and approaches for dealing with the challenges faced in the supply chain whether the challenges are due to unexpected disruptions, planned shutdowns, accidental occurrences, scheduling mismanagement, or deliberate sabotage. Each catastrophic event can often be addressed by the multiple and integrated methodologies of industrial engineering. For decades, industrial engineering has had a practical relationship with what is now popular known as supply chain and logistics. In the early days, what we call logistics now was called "physical distribution" by industrial engineers. Consumers often erroneously think of the supply chain as being represented by the transportation and distribution aspects of providing products, services, and desired results. In fact, the supply chain involves both the upstream and downstream components. In this regard, the production end (the source) of the spectrum is as important as the transportation and distribution components (the destination).

What makes industrial engineering very applicable to the variety of supply chain problems is the fact that the discipline directly confronts who supplies whom with what, when, where, how, and why. The approach covers the spectrum of what can happen and proactively leverages contingency planning. The adaptive definition of industrial engineering presented below bears out this claim.

> *Industrial Engineer* – one who is concerned with the design, installation, and improvement of integrated systems of people, materials, information, equipment, and energy by drawing upon specialized knowledge and skills in the mathematical, physical, and social sciences, together with the principles and methods of engineering analysis and design to specify, predict, and evaluate the results to be obtained from such systems.

This definition embodies the various aspects of what an industrial engineer does. For several decades, the military had called upon the discipline of industrial engineering to achieve program effectiveness and operational efficiencies. Industrial engineering is versatile, flexible, adaptive, and diverse. It can be seen from the definition that a systems orientation permeates the work of industrial engineers. This is particularly of interest to the military because military operations and functions are constituted by linking systems. The major functions of industrial engineers include the following:

- Designing integrated systems of people, technology, process, and methods;
- Modeling operations to optimize cost, quality, and performance trade-offs;

  DOI: 10.1201/9781032620701-2

- Developing performance modeling, measurement, and evaluation for systems;
- Developing and maintain quality standards for government, industry, and business;
- Applying production principles to pursue improvements in service organizations;
- Incorporating technology effectively into work processes;
- Developing cost mitigation, avoidance, or containment strategies;
- Improving overall productivity of integrated systems of people, tools, and processes;
- Planning, organizing, scheduling, and controlling programs and projects;
- Organizing teams to improve efficiency and effectiveness of an organization;
- Installing technology to facilitate workflow;
- Enhancing information flow to facilitate smooth operation of systems;
- Coordinating materials and equipment for effective systems performance;
- Designing work systems to eliminate waste and reduce variability;
- Designing jobs (determining the most economical way to perform work);
- Setting performance standards and benchmarks for quality, quantity, and cost;
- Designing and installing facilities to incorporate human factors and ergonomics.

Industrial engineering (IE) can be described as the practical application of the combination of engineering fields together with the principles of scientific management. It is the engineering of work processes and the application of engineering methods, practices, and knowledge to production and service enterprises. Industrial engineering places a strong emphasis on an understanding of workers and their needs in order to increase and improve production and service activities. Figure 2.1 illustrates how industrial engineering serves as an umbrella discipline for many sub-specialties that apply directly to any complex supply chain. Some of the sub-areas are independent disciplines in their own rights. So coveted are the diverse skills of industrial engineers that they are sought after as process improvement engineers in hospitals, retail organizations, banking, and general financial services sectors.

Industrial engineering has a proud heritage with a link that can be traced back to the *industrial revolution*. Although the practice of industrial engineering has been in existence for centuries, the work of Frederick Taylor in the early 20th century was the first formal emergence of the profession. It has been referred to with different names and connotations. Scientific management was one of the early names used to describe what industrial engineers did.

Industry, the root of the profession's name, clearly explains what the profession is about. The dictionary defines industry generally as the ability to produce and deliver goods and services. The "industry" in industrial engineering can be viewed as the application of skills and creativity to achieve work objectives. This relates to how human effort is harnessed innovatively to carry out work. Thus, any activity can be defined as "industry" because it generates a product, be it service or physical product. A systems view of industrial engineering encompasses all the details and aspects necessary for applying skills and cleverness to produce work efficiently.

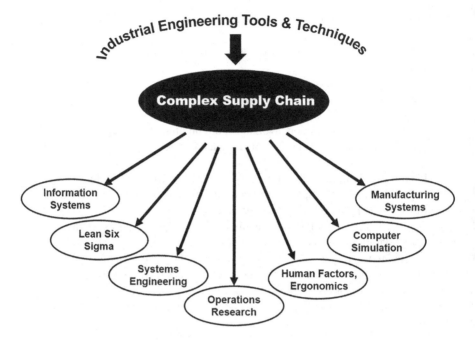

FIGURE 2.1   Industrial engineering and sub-specialties applied to complex supply chains.

The academic curriculum of industrial engineering continues to change, evolve, and adapt to the changing operating environment of the profession.

It is widely recognized that the occupational discipline that has contributed the most to the development of modern society is *engineering*, through its various segments of focus. Engineers design and build infrastructures that sustain the society. These include roads, residential and commercial buildings, bridges, canals, tunnels, communication systems, healthcare facilities, schools, habitats, transportation systems, and factories. Across all of these, the industrial engineering process of systems integration facilitates the success of the efforts. In this sense, the scope of industrial engineering goes through the levels of activity, task, job, project, program, process, system, enterprise, and society. This handbook of military industrial engineering presents essential tools for the levels embodied by this hierarchy of functions. From the age of horse-drawn carriages and steam engines to the present age of intelligent automobiles and aircraft, the impacts of industrial engineering cannot be mistaken, even though the contributions may not be recognized in the context of the conventional work of industry.

Going further back in history, several developments helped form the foundation for what later became known as industrial engineering. In America, George Washington was said to have been fascinated by the design of farm implements on his farm in Mount Vernon. He had an English manufacturer send him a plow built to his specifications that included a mold on which to form new irons when old ones were worn out, or would need repairs. This can be described as one of the earliest attempts to create a process of achieving a system of interchangeable parts. Thomas

Jefferson invented a wooden mold board which, when fastened to a plow, minimized the force required to pull the plow at various working depths. This is an example of early agricultural industry innovation. Jefferson also invented a device that allowed a farmer to seed four rows at a time. In pursuit of higher productivity, he invented a horse-drawn threshing machine that did the work of ten men.

Meanwhile in Europe, productivity growth, through reductions in manpower, marked the technological innovations of the 1769–1800 Europe. Sir Richard Arkwright developed a practical code of factory discipline. In their foundry, Matthew Boulton and James Watt developed a complete and integrated engineering plant to manufacture steam engines. They developed extensive methods of market research, forecasting, plant location planning, machine layout, workflow, machine operating standards, standardization of product components, worker training, division of labor, work study, and other creative approaches to increasing productivity. Charles Babbage, who is credited with the first idea of a computer, documented ideas on scientific methods of managing industry in his book titled *On the Economy of Machinery and Manufacturers*, which was first published in 1832. The book contained ideas on division of labor, paying less for less important tasks, organization charts, and labor relations. These were all forerunners of modern industrial engineering.

In the early history of the United States, several efforts emerged to form the future of the industrial engineering profession. Eli Whitney used mass production techniques to produce muskets for the US Army. In 1798, Whitney developed the idea of having machines make each musket part so that it could be interchangeable with other similar parts. By 1850, the principle of interchangeable parts was widely adopted. It eventually became the basis for modern mass production for assembly lines. It is believed that Eli Whitney's principle of interchangeable parts contributed significantly to the Union victory during the US Civil War. Thus, the early practice of industrial engineering made significant contribution to the military. That heritage has continued until today.

The management attempts to improve productivity prior to 1880 did not consider the human element as an intrinsic factor. However, from 1880 through the first quarter of the 20th century, the works of Frederick W. Taylor, Frank and Lillian Gilbreth, and Henry L. Gantt created a long-lasting impact on productivity growth through consideration of the worker and their environment.

Frederick Winslow Taylor (1856–1915) was born in the Germantown section of Philadelphia to a well-to-do family. At the age of 18, he entered the labor force, having abandoned his admission to Harvard University due to impaired vision. He became an apprentice machinist and pattern-maker in a local machine shop. In 1878, when he was 22, he went to work at the Midvale Steel Works. The economy was in a depressed state at the time. Frederick was employed as a laborer. His superior intellect was very quickly recognized. He was soon advanced to the positions of time clerk, journeyman, lathe operator, gang boss, and foreman of the machine shop. By the age of 31, he was made chief engineer of the company. He attended night school and earned a degree in mechanical engineering in 1883 from Stevens Institute. As a work leader, Taylor faced the following common questions:

1. Which is the best way to do this job?
2. What should constitute a day's work?

These are still questions faced by industrial engineers of today. Taylor set about the task of finding the proper method for doing a given piece of work, instructing the worker in following the method, maintaining standard conditions surrounding the work so that the task could be properly accomplished, and setting a definite time standard and payment of extra wages for doing the task as specified. Taylor later documented his industry management techniques in his book titled *The Principles of Scientific Management*.

The work of Frank and Lillian Gilbreth coincided with the work of Frederick Taylor. In 1895, on his first day on the job as a bricklayer, Frank Gilbreth noticed that the worker assigned to teach him how to lay brick did his work three different ways. The bricklayer was insulted when Frank tried to tell him of his work inconsistencies – when training someone on the job, when performing the job himself, and when speeding up. Frank thought it was essential to find one best way to do work. Many of Frank Gilbreth's ideas were similar to Taylor's ideas. However, Gilbreth outlined procedures for analyzing each step of workflow. Gilbreth made it possible to apply science more precisely in the analysis and design of the work place. Developing *therbligs*, which is a moniker for Gilbreth spelled backwards, as elemental predetermined time units, Frank and Lillian Gilbreth were able to analyze the motions of a worker in performing most factory operations in a maximum of 18 steps. Working as a team, they developed techniques that later became known as work design, methods improvement, work simplification, value engineering, and optimization. Lillian (1878–1972) brought to the engineering profession the concern for human relations. The foundation for establishing the profession of industrial engineering was originated by Frederick Taylor and Frank and Lillian Gilbreth.

The work of Henry Gantt (1861–1919) advanced the management movement from an industrial management perspective. He expanded the scope of managing industrial operations. His concepts emphasized the unique needs of the worker by recommending the following considerations for managing work:

1. Define the task, after a careful study.
2. Teach them how to do it.
3. Provide an incentive in terms of adequate pay or reduced hours.
4. Provide an incentive to surpass it.

Henry Gantt's major contribution is the Gantt chart, which went beyond the works of Frederick Taylor or the Gilbreths. The Gantt chart related every activity in the plant to the factor of time. This was a revolutionary concept for the time. It led to better production planning control and better production control. This involved visualizing the plant as a whole, like one big system made up of interrelated sub-systems. The major chronological historical, scholarly, intellectual, and practical developments marking the applications of industrial engineering are summarized below, albeit not necessarily under the name of "industrial engineering." The essence of the profession is what matters as highlighted in the chronology below, which contains several elements of building, maintaining, and sustaining supply chains in business and industry:

1440: Venetian ships are reconditioned and refitted on an assembly line.

1474: Venetian Senate passes the first patent law and other industrial laws.

1568: Jacques Besson publishes illustrated book on iron machinery as replacement for wooden machines.

1722: Rene de Reaunur publishes the first handbook on iron technology.

1733: John Kay patents the flying shuttle for textile manufacture – a landmark in textile mass production.

1747: Jean Rodolphe Perronet establishes the first engineering school.

1765: Watt invents the separate condenser, which made the steam engine the power source.

1770: James Hargreaves patents his "spinning jenny." Jesse Ramsden devises a practical screw-cutting lathe.

1774: John Wilkinson builds the first horizontal boring machine.

1775: Richard Arkwright patents a mechanized mill in which raw cotton is worked into thread.

1776: James Watt builds the first successful steam engine, which became a practical power source; Adam Smith discusses the division of labor in *The Wealth of Nations*.

1785: Edmund Cartwright patents a power loom.

1793: Eli Whitney invents the "cotton gin" to separate cotton from its seeds.

1797: Robert Owen uses modern labor and personnel management techniques in a spinning plant in the New Lanark Mills in Manchester, England.

1798: Eli Whitney designs muskets with interchangeable parts.

1801: Joseph Marie Jacquard designs automatic control for pattern-weaving looms using punched cards.

1802: "Health and Morals Apprentices Act" in Britain aims at improving standards for young factory workers; Marc Isambard Brunel, Samuel Benton, and Henry Maudsey design an integrated series of 43 machines to mass produce pulley blocks for ships.

1818: Institution of Civil Engineers founded in Britain.

1824: The repeal of the Combination Act in Britain legalizes trade unions.

1829: Mathematician Charles Babbage designs the "analytical engine," a forerunner of the modern digital computer.

1831: Charles Babbage publishes *On the Economy of Machines and Manufacturers*.

1832: The Sadler Report exposes the exploitation of workers and the brutality practiced within factories.

1833: Factory law enacted in United Kingdom. The Factory Act regulates British children's working hours; a general Trades Union is formed in New York.

1835: Andrew Ure publishes *The Philosophy of Manufacturers*, describing the new industrial system that had developed in England over the course of the previous century. He advised that the new factory system is beneficial to workers because it relieved them of much of the tedium of manufacturing goods by hand. It is in this same year that Samuel

Morse invents the telegraph, which opened up communication between remote locations. This invention is the forerunner of digital communication as we know it today.

1845: Friedrich Engels publishes *Condition of the Working Classes in England*, which was a rigorous, formal study of the industrial working class in Victorian England. This same sense of paying attention to the conditions, plight, and opportunities of workers can still be seen in the industrial engineering of today.

1847: The British Government passed the *Factory Act* to improve conditions of women and children working in factories. Young children were working very long hours in deplorable working environments. Under the Act, no child workers under nine years of age were allowed and employers must have an age certificate for their child workers; children aged 9–13 years could work no more than nine hours a day. George Stephenson founded the Institution of Mechanical Engineers. He was a British civil engineer and mechanical engineer. Renowned as the "Father of Railways," Stephenson was considered by the Victorians a great example of diligent application and thirst for improvement. It can be seen that the early works of industrial engineering mirror the early industrial focus of mechanical engineering. The "thirst for improvement" is still seen in today's industrial engineering commitment to continuous improvement.

1948: Frank B. Gilbreth Jr. and Ernestine Gilbreth Carey publish *Cheaper by the Dozen*, which celebrates the professional and family accomplishments of their parents, Frank and Lillian Gilbreth.

1856: Henry Bessemer revolutionizes the steel industry through a novel design for a converter.

1869: Transcontinental railroad completed in United States.

1871: British trade unions are legalized by Act of Parliament.

1876: Alexander Graham Bell invents a usable telephone. As an advancement over Samuel Morse's telegraph, the usable telephone facilitated and expedited communication beyond comprehension in that era.

1877: Thomas Edison invents the phonograph. The phonograph was developed as a result of Thomas Edison's work on two other inventions, the telegraph and the telephone. In 1877, Edison was working on a machine that would transcribe telegraphic messages through indentations on paper tape, which could later be sent over the telegraph repeatedly. The recorded voice quickly became a tool for operational improvements in industrial establishments.

1878: Frederick W. Taylor joins Midvale Steel Company. Taylor was widely known for his methods to improve industrial efficiency. He was one of the first management consultants. Taylor was one of the intellectual leaders of the Efficiency Movement and his ideas, broadly conceived, were highly influential in the Progressive Era of the 1890s to 1920s. Efficiency, in any form, whether digital or analog, has remained a focus in the practice of industrial engineering of today.

1880: American Society of Mechanical Engineers (ASME) is organized.

1881: Frederick Taylor begins time study experiments that would lay the foundation for the early introduction of industrial engineering to industries.

1885: Frank B. Gilbreth begins motion study research. A *time and motion study* (or *time-motion study*) is an efficiency technique that combined the time study work of Frederick Taylor with the motion study work of Frank and Lillian Gilbreth. It was a major part of scientific management (also known as Taylorism). After its first introduction, time study developed in the direction of establishing standard times for work elements, while motion study evolved into a technique for improving work methods. The two techniques became integrated and refined into a widely accepted method applicable to the improvement and upgrading of work systems, which is still practiced by industrial engineering today under the name of methods engineering, which can be found in industrial establishments, service organizations, banks, schools, hospitals, government, and the military. The tools and techniques of industrial engineering are everywhere.

1886: Henry R. Towne presents the paper "The Engineer as Economist."; American Federation of Labor (AFL) is organized; Vilfredo Pareto publishes *Course in Political Economy*; Charles M. Hall and Paul L. Herault independently invent an inexpensive method of making aluminum.

1888: Nikola Tesla invents the alternating current induction motor, enabling electricity to take over from steam as the main provider of power for industrial machines; Dr. Herman Hollerith invents the electric tabulator machine, the first successful data processing machine.

1890: Sherman Anti-Trust Act is enacted in the United States.

1892: Gilbreth completes motion study of bricklaying.

1893: Taylor begins work as consulting engineer.

1895: Taylor presents paper titled "A Piece-Rate System" to ASME.

1898: Taylor begins time study at Bethlehem Steel; Taylor and Maunsel White develop process for heat-treating high-speed tool steels.

1899: Carl G. Barth invents a slide rule for calculating metal cutting speed as part of Taylor system of management.

1901: American national standards are established; Yawata Steel begins operation in Japan.

1903: Taylor presents paper titled "Shop Management" to ASME; H.L. Gantt develops the "Gantt chart"; Hugo Diemers writes *Factory Organization and Administration*; Ford Motor Company is established.

1904: Harrington Emerson implements Santa Fe Railroad improvement; Thorstein B. Veblen publishes *The Theory of Business Enterprise*.

1906: Taylor establishes metal-cutting theory for machine tools; Vilfredo Pareto publishes *Manual of Political Economy*.

1907: Frank Gilbreth uses time study for construction.

1908: Model T Ford is built; Pennsylvania State College introduces the first university course in industrial engineering.

1911: Frederick Taylor publishes *The Principles of Scientific Management*; Frank Gilbreth publishes *Motion Study*; factory laws are enacted in Japan.

1912: Harrington Emerson publishes *The Twelve Principles of Efficiency*; Frank and Lillian Gilbreth presented the concept of "therbligs"; Yokokawa translates into Japanese Taylor's *Shop Management* and *The Principles of Scientific Management*.

1913: Henry Ford establishes a plant at Highland Park, Michigan, which utilizes the principles of uniformity and interchangeability of parts, and of the moving assembly line by means of conveyor belt; Hugo Munstenberg publishes *Psychology of Industrial Efficiency*.

1914: World War I starts; Clarence B. Thompson edits *Scientific Management*, a collection of articles on Taylor's system of management.

1915: Taylor's system is used at Niigata Engineering's Kamata plant in Japan; Robert Hoxie publishes *Scientific Management and Labour*; Lillian Gilbreth earned PhD at Brown University in 1915 in psychology.

1916: Lillian Gilbreth publishes *The Psychology of Management*; Taylor Society established in United States.

1917: Frank and Lillian Gilbreth publish *Applied Motion Study*; the Society of Industrial Engineers is formed in the United States.

1918: Mary P. Follet publishes *The New State: Group Organization, the Solution of Popular Government*.

1919: Henry L. Gantt publishes *Organization for Work*.

1920: Merrick Hathaway presents paper "Time Study as a Basis for Rate Setting"; General Electric establishes divisional organization; Karel Capek introduces *Rossum's Universal Robots*. This play coined the word "robot."

1921: The Gilbreths introduce process-analysis symbols to ASME (American Society of Mechanical Engineers).

1922: Toyoda Sakiichi's automatic loom is developed; Henry Ford publishes *My Life and Work*.

1924: Frank and Lillian Gilbreth announce results of micromotion study using therbligs; Elton Mayo conducts illumination experiments at Western Electric.

1926: Henry Ford publishes *Today and Tomorrow*.

1927: Elton Mayo and others begin relay-assembly test room study at the Hawthorne plant.

1929: Great Depression; International Scientific Management Conference held in France.

1930: Hathaway publishes *Machining and Standard Times*; Allan H. Mogensen discusses 11 principles for work simplification in *Work Simplification*; Henry Ford publishes *Moving Forward*.

1931: Dr. Walter Shewhart publishes *Economic Control of the Quality of Manufactured Product*.

1932: Aldous Huxley publishes *Brave New World*, the satire which prophesies a horrifying future ruled by industry.

1934: General Electric performs micromotion studies.

1936: The word "automation" is first used by D.S. Harder of General Motors. It is used to signify the use of transfer machines which carry parts automatically from one machine to the next, thereby linking the tools into an integrated production line. Charlie Chaplin produces *Modern Times*, a film showing an assembly line worker driven insane by routine and unrelenting pressure of his job.

1937: Ralph M. Barnes publishes *Motion and Time Study*.

1941: R.L. Morrow publishes "Ratio Delay Study," an article in *Mechanical Engineering* journal; Fritz J. Roethlisberger publishes *Management and Morale*.

1943: ASME work standardization committee publishes glossary of industrial engineering terms.

1945: Marvin E. Mundel devises "memo-motion" study, a form of work measurement using time-lapse photography; Joseph H. Quick devises work factors (WF) method; Shigeo Shingo presents concept of production as a network of processes and operations and identifies lot delays as source of delay between processes, at a technical meeting of the Japan Management Association.

1946: The first all-electronic digital computer ENIAC (Electronic Numerical Integrator and Computer) is built at Pennsylvania University; The first fully automatic system of assembly is applied at the Ford Motor Plant.

1947: American mathematician Norbert Wiener publishes *Cybernetics*.

1948: H.B. Maynard and others introduce methods time measurement (MTM) method; Larry T. Miles develops value analysis (VA) at General Electric; Shigeo Shingo announces process-based machine layout; American Institute of Industrial Engineers is formed.

1950: Marvin E. Mundel publishes *Motion and Time Study, Improving Productivity*.

1951: Inductive statistical quality control is introduced to Japan from the United States.

1952: Role and sampling study of industrial engineering conducted at ASME.

1953: B.F. Skinner publishes *Science of Human Behaviour*.

1956: New definition of industrial engineering is presented at the American Institute of Industrial Engineers Convention: "Industrial engineering is concerned with the design, improvement and installation of integrated systems of people, material, information, equipment and energy. It draws upon specialized knowledge and skills in the mathematical, physical and social sciences, together with the principles and methods of engineering analysis and design to specify, predict and evaluate the results to be obtained from such systems."

1957: Several books are published, including Chris Argyris, *Personality and Organization*; Herbert A. Simon, *Organizations*; and R.L. Morrow, *Motion and Time Study*. Shigeo Shingo introduces scientific thinking mechanism (STM) for improvements; the Treaty of Rome established the European Economic Community.

1960:   Douglas M. McGregor publishes *The Human Side of Enterprise.*
1961:   Rensis Lickert publishes *New Patterns of Management*; Shigeo Shingo devises ZQC (source inspection and poka-yoke systems); Texas Instruments patents the silicon chip integrated circuit.
1963:   H.B. Maynard publishes *Industrial Engineering Handbook*; Gerald Nadler publishes *Work Design.*
1964:   Abraham Maslow publishes *Motivation and Personality.*
1965:   Transistors are fitted into miniaturized "integrated circuits."
1966:   Frederick Hertzberg publishes *Work and the Nature of Man.*
1967:   Sidney Gordon Gilbreath III completes PhD dissertation on *A Bayesian Procedure for the Design of Sequential Sampling Plans* at Georgia Institute of Technology.
1968:   Roethlisberger publishes *Man in Organizations*; Department of Defense publishes *Principles and Applications of Value Engineering.*
1969:   Shigeo Shingo develops single-minute exchange of dies (SMED); Shigeo Shingo introduces pre-automation; Wickham Skinner publishes the article "Manufacturing – Missing Link in Corporate Strategy" in *Harvard Business Review.*
1971:   Taiichi Ohno completes the Toyota production system; Intel Corporation develops the micro-processor chip.
1973:   First annual Systems Engineering Conference of AIIE.
1975:   Shigeo Shingo extols NSP-SS (non-stock production) system; Joseph Orlicky publishes *MRP: Material Requirements Planning.*
1976:   IBM markets the first personal computer.
1980:   Matsushita Electric used Mikuni method for washing machine production; Shigeo Shingo publishes *Study of the Toyota Production System from an Industrial Engineering Viewpoint.*
1981:   Oliver Wight publishes *Manufacturing Resource Planning (MRP II)*; AIIE (American Institute of Industrial Engineers) became IIE (Institute of Industrial Engineers).
1982:   Gavriel Salvendy publishes *Handbook of Industrial Engineering.*
1984:   Shigeo Shingo publishes *A Revolution in Manufacturing* publishes *The SMED System.* Emergence of more formal practice area named Hospital Industrial Engineering to leverage industrial engineering tools and techniques for operational improvement in the healthcare industry. National hospital operational excellence has since been credited to the application of industrial engineering.
1989:   Development of Code Division Multiple Access (CDMA) for cellular communications.
1990:   Wide use of the concept of Total Quality Management (TQM); Kjell Zandin publishes *MOST Work Measurement Systems: Basic Most, Mini Most, Maxi Most.*
1993:   Adedeji B. Badiru and B. J. Ayeni publish *Practitioner's Guide to Quality and Process Improvement.*
1995:   Adedeji B. Badiru publishes *Industry's Guide to ISO 9000*; The dot-com boom started in earnest; Netscape search engine was introduced; Peter

Norvig and Stuart Norvig publish *Artificial Intelligence: A Modern Approach*, which later became the authoritative textbook on AI.

2000: The turning point of the 21st century and the Y2K computer date scare.

2004: The birth of Facebook social networking; Skype took over worldwide online communication.

2008: The National Academy of Engineering (NAE) publishes the 14 Grand Challenges for Engineering; Adedeji B. Badiru et al. publish *Industrial Project Management: Concepts, Tools, and Techniques.*

2009: Adedeji B. Badiru and Marlin Thomas publish the *Handbook of Military Industrial Engineering* to promote the application of industrial engineering in national defense strategies, which won the won the 2010 Book-of-the-Year Award from IISE.

2012: Adedeji B. Badiru et al. publish *Industrial Control Systems: Mathematical and Statistical Models and Techniques.*

2014: Adedeji B. Badiru publishes the second edition of the Handbook of Industrial and Systems Engineering; unmanned aerial vehicles (UAVs, or drones) emerged as practical for a variety of applications.

2016: IIE (Institute of Industrial Engineers) changed name to IISE (Institute of Industrial & Systems Engineering); Wide appearance of self-driving cars; Adedeji B. Badiru publishes *Global Manufacturing Technology Transfer.*

2017: Adedeji B. Badiru and Sharon Bommer publish *Work Design: A Systematic Approach*; A. Ravi Ravindran and Donald Warsing publish *Supply Chain Engineering: Models and Applications*; Internet of Things (IOT) makes a big splash.

2018: Emergence of hybrid academic programs encompassing industrial engineering, digital engineering, data analytics, and virtual reality simulation.

2019: Adedeji B. Badiru publishes *The Story of Industrial Engineering*, which won the 2020 Book-of-the-Year award from IISE; Adedeji B. Badiru and Cassie Barlow edit *Defense Innovation Handbook: Guidelines, Strategies, and Techniques*; Adedeji B. Badiru publishes *Systems Engineering Models: Theory, Methods, and Applications*; Adedeji B. Badiru et al. publish *Manufacturing and Enterprise: An Integrated Systems Approach.*

2020: Tools and techniques of industrial engineering applied to supply chain networks related to healthcare emergency needs necessitated by the COVID-19 pandemic; Adedeji B. Badiru publishes *Innovation: A Systems Approach.*

2021: Adedeji B. Badiru and Lauralee Cromarty publish *Operational Excellence in the new Digital Era*; Adedeji B. Badiru and Tina Agustiady publish *Sustainability: A Systems Engineering Approach to the Global Grand Challenge*; Adedeji B. Badiru publishes *Data Analytics: Handbook of Formulas and Techniques.*

2022: Adedeji B. Badiru publishes *Global Supply Chain: Using Systems Engineering Strategies to Respond to Disruptions.*

As can be seen from the historical details above, industrial engineering has undergone progressive transformation over the past several decades and its applicability to the supply chain system is not in question. Frank and Lillian Gilbreth, whose names appear several times in the chronological accounts, were industrial engineers. They were among the first in the scientific field of operations management and the first in motion study and analysis. From 1910 to 1924, their company, Gilbreth, Inc., was employed as "efficiency experts" by many of the major industrial plants in the United States, Britain, and Germany. When Frank Gilbreth died in 1924, his wife, Lillian, carried on the work and became, perhaps, the foremost woman industrial engineer. She remained professionally active until her death in 1972.

PRODUCTION, QUALITY INSPECTION, AND THE SUPPLY CHAIN

Any supply chain relies on the production of products that the market wants. The shipment of finished goods is under the assumption that the products destined for shipment (through the supply chain) are of acceptable quality. Rejected quality ultimately affects the ability of the producer to meet market demands for the products with a reasonable time span. As long ago as 1976, S. G. Gilbreath (Gilbreath, 1967) recognized this linkage and did an extensive doctoral research on acceptance sampling. Although industry has been moving away from acceptance sampling under the pretext of focusing on using Lean Six Sigma techniques to produce products of acceptable quality, Gilbreath's foundational research still offers some seminal insights into the classical role of quality inspection at the source. With good inspection practices, more acceptable products can be generated to keep the supply chain humming well, even when there are unexpected disruptions, such as the COVID-19 pandemic. An excerpt of Gilbreath's research results is presented here as a guide, hopefully, for future researchers interested in linking quality inspection to the robustness of modern supply chains. There are ideas and leads for contemporary research studies in the context of the prevailing global supply chains, driven by modern interconnectedness of producers, markets, and shipping services.

In Gilbreath's research (Gilbreath, 1967 and references therein), mathematical models for item-by-item sequential sampling of attributes are developed. The models consider prior distributions of process quality, inspection costs, and decision losses. Models are presented in which the prior distributions are:

1. Hypergeometric
2. Binomial
3. Polya
4. Mixed binomial.

Bayesian decision rules are the bases for computing expected losses. A conceptual model for identifying and measuring inspection costs and decision losses is formulated. This model is applied to a hypothetical example (Gilbreath, 1967). The work of previous investigators of sampling procedures who have considered prior distributions and costs is extended in four ways. These are:

1. Development of a concept for identifying and measuring inspection costs and decision losses;
2. Design of general cost models for item-by-item sequential sampling of attributes considering prior distributions, inspection costs, and decision losses;
3. Design of a computer program of the sampling models in order that they be readily adaptable to inspection procedures using online access data processing equipment;
4. Analysis of the sensitivity of the sampling models to changes in decision losses.

The difficulty in identifying and measuring relevant cost criteria for sampling inspection decisions may, in many cases, be attributed to attempts by researchers to consider too broad a spectrum of possible applications. By first outlining a procedure for establishing a model of a specific enterprise's situation and then analyzing that particular model, the determination and allocation of economic criteria are rendered far less formidable. If relatively accurate criteria can be obtained, they are, in the opinion of many researchers of sampling procedures, superior to statistical criteria as the bases for designing or choosing sampling inspection procedures. The relevant economic criteria to be considered are:

1. C_f, the fixed sampling cost;
2. C_s, the variable sampling cost, per item inspected;
3. C_a, the decision loss accompanying acceptance of a defective item;
4. C_r, the decision loss resulting from rejection of a good item.

The decision to be made at any point in the course of inspection is among the alternatives:

1. Accept the lot now;
2. Reject the lot now;
3. Inspect one more item and accept or reject the lot.

The alternative decision carrying the lowest total expected cost, where total expected cost is the sum of inspection costs and decision losses is selected. If (1) or (2) is most economical, inspection is terminated. If (3) is most economical, one more item is inspected. Although (3) is defined for purposes of cost evaluation as "inspect one more item and accept or reject the lot," actual acceptance or rejection is deferred until (1), (2), and (3) are re-evaluated for each new sample. Inspection continues until at some point decision (1) or (2) is made and the lot is categorized appropriately. N is the lot size, n is the accumulative sample size, x is the cumulative number of defective items in the accumulated sample, y is the number of defectives in $(N - n)$, the uninspected portion of a lot, and is the probability of acceptance or is the probability that the next, $(n + 1)$st, item inspected is good given that x defective items have been found in a cumulative sample of n items. X is the number of defective items in a lot.

GILBREATH'S DECISION MODEL

$$\text{For}: \quad n = (0, 1, ..., N),$$

$$x = [\text{Max.} \ (0, n - N + x), ... \text{Min.} \ (n, x)]$$

$$X = (0, 1, ..., N), \text{ and}$$

$$y = (X - x)$$

The expected costs and losses of the three alternative decisions are:

1. $C_a E(y|x_n)$, where the subscript n indicates the result corresponds to a cumulative sample of size n,

2. $C_r[(N - n) - E(y|x_n)]$

3. $C_f + Cs + Ca \ E[y|(x = 0)_{n=1}]P_A$

   $+ c_r\{(N-1) - E[y|(x = 1)_{n=1}]\}(1 - P_A)$, for n = 0

   $C_{s+}C_a E(y|X_{n+1}) \ P_A + C_r[N - n - 1) - E(y|x + 1_{n+1})] \ (1 - P_A)$

   for n = (1, 2, ..., N)

At any value of n when either (1) or (2) becomes cheaper than (3), inspection ceases and the lot is categorized appropriately. This iterative procedure is computer programmed to adapt the sequential plans to online access equipment. The sensitivity analysis consists of demonstrating the response of the proposed model to shifts in the relevant statistical and economic parameters. Several specific cases are examined.

The sequential procedure was compared to categorizing without inspection and to a system of optimum single sampling plans for the two-point mixed binomial prior distribution over a variety of statistical and economic parameters. The proposed plans were in no case less economical than categorizing without inspection. For all conditions tested the proposed procedure was significantly inferior to the optimum single sampling plans. Sensitivity of the sequential model to changes in quality levels of submitted lots and to changes in decision losses was favorable under all conditions tested.

ACCEPTANCE SAMPLING FOR QUALITY CONTROL

The purpose of Gilbreath's research (Gilbreath, 1967) is to contribute to industrial understanding of and ability to use item-by-item sequential sampling. This is accomplished by providing a model, whereby in practice, item-by-item sequential sampling plans can be designed when certain costs are known and a specific prior distribution of product quality exists. A conceptual model for determining relevant costs and losses is presented.

The objective of Gilbreath's study is to develop a mathematical model which will specify an item-by-item sequential sampling plan to be used in a given situation. Samples will be considered to be drawn from lots selected randomly from a process thereby generating a probability distribution of product quality. Hald's compound hypergeometric distribution (Gilbreath, 1967) describes the sampling procedure. The specific objectives of Gilbreath's research are:

1. Development of a concept for allocating inspection costs and decision losses on a per-occurrence basis;
2. Design of cost models for sequential sampling of attributes considering prior distributions, inspection costs, and decision losses;
3. Comparison of the sequential plans developed herein with the optimum single plans presented by A. Hald (Gilbreath, 1967);
4. Computer programming of the sequential sampling model to demonstrate the adaptability of the model to online access data processing equipment;
5. Analysis of the sensitivity of the sequential sampling models to shifts in allocated decision losses.

Acceptance sampling in statistical quality control has been the subject of much effort and scientific inquiry since industrialists, quality control practitioners, and statisticians first became aware that, under certain circumstances, it is cheaper to inspect samples of product and "take the chance" that the sample is truly representative of the lot than to examine every item in the lot (Gilbreath, 1967 and all references therein). Many rules have been established for determining sample sizes and acceptance or rejection criteria. There are almost as many bases for such rules as there are rules themselves. There are plans that give a certain "assurance" of accepting lots, whose quality is better than some minimum desirable level. There are plans that give another certain "assurance" that lots worse than some maximum desirable quality level will be rejected. Other plans attempt to give both these assurances and, in most cases because the sampling distributions are discrete, result in some minor compromise in respect to at least one of these assurances.

There are many formulas for finding the expected costs of such sampling plans, and in specific cases they may be helpful in determining whether it is cheaper to inspect 100%, to sample, or not to inspect at all. There is a serious limitation in most of these cost determinations, however. They depend upon some prior knowledge (or guess) concerning the true lot product quality. That is to say, such cost determinations are based upon conditional probability, the probability of lot acceptance given a specific lot quality.

In 1960, A. Hald, at the University of Copenhagen (per Gilbreath, 1967), based single sampling plans upon a model more closely resembling reality. Industrial processes tend to produce quality within some process capability. As long as this process capability is the result of a constant cause system, product quality will generate a probability distribution. The probability that there are X defectives in a lot $= p(X)$. Furthermore, if there are N items in a lot, X may vary from 0 to N or, $0 < X < N$. When a sample of n items is taken from the lot, the probability that the sample contains x defective items $= p(x|x)$ where $0 < x < n$ and $0 < x < X$. In other words, if only this last probability is considered, then some assumption must be made about X, the number of defectives in the lot. This is the approach taken by investigators prior to the work of Hald. By considering p(X), the prior distribution, Hald approached more closely the true sampling situation, that the probability of x defectives in a sample equals the probability of x defectives in a sample given all possible values of X defectives in the lot from which the sample was drawn. Letting x = number of defectives in the sample and y = number of defectives in the remainder of the lot,

$$X = x + y$$

and

$$p\,(x, y) = P_N\,(x + y)\begin{bmatrix} n \\ x \end{bmatrix}\begin{bmatrix} N - n \\ y \end{bmatrix} / \begin{bmatrix} N \\ x + y \end{bmatrix}$$

from which Hald derives the marginal distribution of x:

$$g_N\,(x) = \begin{bmatrix} n \\ x \end{bmatrix}\sum\nolimits_{y=0}^{N-n} PN(x + y)\begin{bmatrix} N - n \\ y \end{bmatrix} / \begin{bmatrix} N \\ x + y \end{bmatrix}$$

The costs and losses to be considered are those contributing to product cost and related to product quality. They are:

1. The loss incurred by accepting bad lots;
2. The loss incurred by rejecting good lots;
3. The costs of inspecting items.

An inspection procedure that minimizes an appropriate function of these costs is desired when it is appropriate to select an inspection procedure. If all criteria are reduced to the three economic ones above, a model or models to minimize the total cost function would be universally applicable. According to Gilbreath (1967), Hald has formulated such models for single sampling. At the conclusion of his work, he suggested that his models be extended so as to apply to sequential sampling. It is toward such an extension that the proposed research is directed.

If the objectives of this research are accomplished, the resulting sequential sampling model, together with Hald's models, will provide economic bases for selecting single or sequential sampling plans. This would constitute a large and practical step forward, toward reality, from existing sampling plan selection techniques based upon somewhat nebulous probabilistic considerations related to economics only through implication.

It is the intent of Gilbreath's research to extend the then current practice in item-by-item sequential sampling to include consideration of prior distribution and costs. The cost ratios used by Hald in his single sampling work is investigated relative to their extended applicability to sequential analysis. The binomial, hypergeometric, Polya, and mixed binomial prior distributions investigated by Hald, Johnson, and Schafer (see Gilbreath, 1967) will be applied to sequential sampling practice. A Bayesian statistical model will be designed, whereby sequential sampling plans may be designed considering prior distribution and using cost as the measure of effectiveness.

Gilbreath's study considers item-by-item sequential sampling for the attributes, good or bad. Acceptance sampling by attributes has been studied carefully to develop an understanding of the strengths and weaknesses of then existing procedures for designing acceptance sampling plans. Consideration is given design schemes which do not depend explicitly upon economic criteria and others which do. Careful attention is given to a study of the state of the art (science) of the identification and measurement of costs and losses comprising relevant criteria for acceptance sampling decisions. Further investigations have been made into the usefulness of existing criteria and measurement techniques. The study is limited to product inspection where the inspection is performed for the purpose of accepting lots of products containing few defective items and rejecting those showing indications of containing many. No attention was given variables inspection designs. No consideration is given to the psychological aspects of acceptance sampling. This may be a good lead for modern research focusing on human factors and ergonomics in industrial settings. Gilbreath's research is confined to the economic, statistical, and mathematical theories of acceptance sampling for attributes.

The inspection of product could occur in a wide variety of circumstances. The product might be incoming purchased material, outgoing product undergoing final inspection, or product at any stage within a manufacturing process. It is assumed that the process responsible for the product's characteristics up until it is inspected is stable to the degree that a process curve can exist. By process curve is meant f(p), the probability density function for the fraction defective, p.

Gilbreath (1967) presents a comprehensive survey of acceptance sampling. Acceptance sampling literature was reviewed and classified according to the content of each publication. The material studied was primarily classified according to the criteria employed by decision rules. These criteria are either statistical or economic. Statistical criteria employed in designing acceptance sampling plans are similar to those used in significance testing. The risks or probabilities of commission of certain errors are fixed, and sample sizes are determined along with acceptance and rejection criteria to achieve these risks.

SINGLE, DOUBLE, AND MULTIPLE SAMPLING

Acceptance sampling by attributes has been discussed extensively. However, most of what has been written is not pertinent to this research, which deals only with item-by-item sequential sampling by attributes. In order to develop the present status of sequential sampling by attributes, it is helpful to discuss attribute sampling generally. To do this requires mention of single, double, and multiple sampling plans. The most widely used sampling plans are probably those of Military Standard 105 and of Dodge-Romig plans (Gilbreath, 1967).

Military Standard 105: Military Standard 105 presents sampling plans that specify an acceptable quality level (AQL): "the maximum percent defective (or the maximum number of defects per hundred units) that, for purposes of sampling inspection, can be considered satisfactory as a process average." These plans represent significance tests of the hypothesis, percent defective, p′, is equal to AQL against the alternate hypothesis, p′ is greater than AQL. The maximum number of defects allowed in a sample, C, is established, and the sample size, n, is then determined such that the probability of accepting a lot having p′ equal to AQL is equal to $1 - \alpha$. Thus, α is the risk of committing an error of type I as in any significance test. No consideration is given to explicitly designating the risk, and sampling proceeds on this basis. From each lot of N items, examine a sample of n items. If the sample contains C or fewer defectives, accept the lot; otherwise, reject the lot. The characteristic, α, is not constant in the plans of Military Standard 105-D, but varies with lot size and with the inspection level selected. Matched double and multiple sampling plans are specified in 105-D.

Dodge-Romig: In the case of Dodge-Romig plans, the choice of plans is made similarly, except that a constant risk is specified and no consideration is given to α. β, the probability of accepting a lot having $p' =$ lot tolerance percent defective (LTPD) is fixed, the acceptance number C is chosen, and n is determined such that there is a probability of 0.10 of accepting a lot having p' equal to LTPD, and average total inspection is minimized. Effective use of Dodge-Romig acceptance sampling procedures requires that rejected lots be screened and all defectives in the sample and in the screened lots be replaced by good items. Other Dodge-Romig plans are provided whereby the average outgoing quality limit (AOQL), "the maximum of the average outgoing qualities for all possible incoming qualities for a given acceptance sampling plan," is fixed. For both LTPD and AOQL plans, Dodge-Romig plans (Gilbreath, 1967) use minimum average total inspection (ATI) as the criterion for determining a specific plan. Here is a summary guide for Dodge-Romig with reference to economy:

> In choosing a value of LTPD (or AOQL) consider and compare the cost of inspection with the economic loss that would ensue if quality as bad as the LTPD were accepted often (or if the average level of quality were greater than the AOQL). Even though the evaluation of economic loss may be difficult, relative values for different levels of percent defective may often be determined.

Once LTPD is determined, with β fixed at 0.10, "only the least amount of inspection which will accomplish the purpose can be justified." The Dodge-Romig plans are designed to provide minimum ATI and to assure that B is 0.10 when pT is equal to LTPD. Therefore, if β is 0.10 and LTPD were chosen to minimize the costs of accepting bad items and of rejecting good ones, Dodge-Romig plans would be least cost plans. No latitude is allowed in selecting β, and no explicit provision is made for minimizing losses by the choice of LTPD (or AOQL). It must be concluded that, in general, Dodge-Romig plans do not provide least-cost sampling plans. Thus, the plans of Military Standard 105-D assure a high probability of accepting lots of quality AQL or better, and the plans of Dodge-Romig establish a low probability, 0.10, of accepting lots of quality LTPD or poorer. Neither group of plans considers sampling costs and losses explicitly, and neither deals with α and β simultaneously.

Plans specifying both α and β: In 1952 J. M. Cameron (Gilbreath, 1967) developed single and double sampling plans using the Poisson approximation to the binomial (the assumed sampling distribution) which established C and n when acceptable quality, p'_1, unacceptable quality, p'_2, α, and β were specified. Once again, the criteria for test design were α and β, and the design was consistent with Neyman and Pearson's hypothesis testing (Gilbreath, 1967). This approach uses a ratio, p_2/p_1, to establish C. When C is known (there are usually two possibilities for C), there may be four plans which approach the desired operating characteristic. For each of the two C values, the designer must adhere strictly to either α or β, whichever is selected, determine n, and allow the other risk to vary slightly. Then the "best" of the four plans as judged by the designer must be chosen for use. Economy may enter into the choice of the best of the four plans, but economy did not enter into the selection of the four plans in the beginning. In 1964, L. A. Johnson (Gilbreath, 1967) surveyed sampling procedures based upon economic and non-economic criteria, with the following conclusions:

1. Types of inspection schemes presently available provide adequate choice for the form of an acceptance inspection decision rule.
2. Further development of sampling tables based on non-economic criteria would be of little value.
3. Efforts to utilize economic criteria, and knowledge of the process curve to develop inspection procedures, have not resulted in any generally applicable theory.
4. There is no general agreement as to the measure of effectiveness for acceptance inspection operations.
5. There is no general agreement as to the appropriate principle of choice in selecting among alternative inspection procedures.
6. Inspection has not been treated adequately as a system of interrelated operations.
7. Successful analysis and design of inspection systems will require a better developed and more clearly understood set of principles – economic, statistical, mathematical, psychological – than now exists relative to acceptance inspection.

After considering the statistical character of acceptance sampling and various measures of the effectiveness of various procedures, it was concluded (Gilbreath, 1967) that monetary units are the superior measures of effectiveness and that Bayes' principle furnishes the best criterion for acceptance inspection decisions. These conclusions are consistent with A. Hald's work as mentioned previously.

Bayesian decision rules: In research at Western Reserve University, Ray E. Schafer (Gilbreath, 1967) agrees that Bayesian statistics provide superior decision rules for categorizing lots in acceptance sampling. However, he argues that the costs relevant to measuring the effectiveness of sampling inspection are so obscure as to be almost unobtainable. Therefore, statistical criteria must be used as measures of effectiveness. Schafer designed single sampling plans by attributes based upon the posterior producer's and consumer's risks for uniform, binomial, hypergeometric, and Polya prior distributions. These are the same prior distributions considered by Hald and Johnson (Gilbreath, 1967). Schafer defines the posterior producer's risk as, "α^*, the posterior conditional probability that, given lot rejection, the number of defective units in the lot was A such that $z \leq z_1$," and the posterior consumer's risk as, "β^*, the posterior conditional probability that, given lot acceptance, the number of defective units, y, in the uninspected portion of the lot is such that $y \geq y2^*$."

$z_1 = $ A particular value of z such that the producer desires few rejected lots to have $z \leq z1$

y_2 = A particular value of y such that the consumer desires
few accepted lots to have y $\geq y_2$,
$z = x + y$ the total number of defectives in a lot of size N.

According to Gilbreath (1967), Hald, Johnson, and Schafer agree that the prior distribution is of utmost importance to the choice of a sampling procedure. Furthermore, they agree that Bayes' principle is the most intelligent principle of choice.

Sequential sampling: In single sampling for attributes, lots of N items are selected for inspection. A sample of n items is drawn from the lot and x, the number of unfavorable attributes in the sample are counted. If x is equal to or less than c, the maximum number of allowable unfavorable attributes for acceptance, the lot is accepted. If x exceeds c, the lot is rejected. The average number of items inspected per lot is a constant. This causes the average cost of sampling inspection per lot to be a constant, regardless of lot quality. Double and multiple sampling provide for reduced average sample sizes. In effect, these plans divide the total sample size into subsamples, each with its own acceptance number. Acceptance or rejection may occur on any sub-sample. Therefore, lots of particularly good or poor quality have a high probability that a decision will be reached on an early sub-sample. Furthermore, if the acceptance number for a subsample is exceeded during inspection of any sub-sample after the first one, rejection may occur without completing inspection of that subsample. This curtailment of inspection further reduces the inspection required, thereby resulting in an additional reduction in average sampling inspection cost. Item-by-item sequential sampling by attributes gives the opportunity to take the maximum advantage of diminished sampling inspection for lots of very good quality or of very poor quality.

Economic criteria: The use of economic criteria in designing acceptance sampling plans involves the identification of relevant costs and losses, the inclusion of these economic parameters in a sampling model, and the formulation of a decision rule that attempts to minimize the total expected cost of sampling inspection. Costs associated with inspection and decisions should be the design criteria for sampling inspection plans. There is a difficulty of obtaining accurate costs for such conditions as acceptance of bad lots and rejection of good ones, but nevertheless these are the losses, along with inspection costs, producers and consumers are trying to reduce by sampling. Three cost affiliated with sampling inspection are identified:

1. $\dfrac{\text{Average cost of accepting a defective item}}{\text{Average cost of accepting a defective item}} = 1$

2. Case I, sorting, $Kr = \dfrac{Sorting\ cost/item}{Average\ cost\ of\ accepting\ a\ defective\ item}$

   Case II, non-sorting, $Kr = \dfrac{Manufactureing\ cost/item}{Average\ cost\ of\ accepting\ a\ defective\ item}$

3. $ks = \dfrac{Sampling\ and\ testing\ cost\ per\ item\ (In\ Sample)}{Average\ cost\ of\ accepting\ a\ defective\ item}$

APPLICATION OF DECISION THEORY

The literature contains many ideas about the design of sampling procedures using statistical decision theory. The sampling inspection of mass-produced items is a field of special application for the theory of statistical decisions, which has received such an impetus from the work of many researchers. In the theory of statistical decisions, we envisage a situation where the following occur:

1. A choice must be made from among a series of possible actions, on the basis of data subject to random variations.
2. It is possible to associate a definite measure of loss to each possible action when it is inappropriately taken.
3. The choice in question can legitimately be regarded as one of a series of statistically independent choices, so that it is reasonable to act in each case in such a way as to minimize the expected value of the loss.

Situations of this sort rarely occur in practical life. One or more of the three essential features are usually missing. It may not be possible to enumerate clearly in advance all the alternative courses of action; or it may not be possible to associate each alternative with a definite measure of loss. Alternatively, the choice in question may have a unique character which makes it inappropriate to regard it as one of a series. But in the case of sampling inspection of batches of items we do little damage to the true nature of the practical problem if we regard it as coming within the scope of the theory. The alternative courses of action, acceptance of the batch, rejection, acceptance subject to further inspection, downgrading, and so on are fairly clear cut; quite good estimates can often be made of the monetary losses arising from wrong decisions; and routine cases by definition, can be regarded as forming a series, and each decision can be regarded as final.

There are many problems associated with the use of cost criteria and Bayesian decision rules in sampling inspection. Among those are:

1. The importance of the form of the prior distribution;
2. Mood's theorem concerning the pure binomial prior distribution;
3. Evidence for the mixed binomial prior distribution;
4. The dependence of sample size on lot or batch size;
5. Applicability of other principles of choice.

Knowledge of the true form of the process curve or prior distribution is very important when applying statistical decision theory in sampling inspection. Gilbreath (1967) echoed G. E. P. Box's comment that it is usually a relatively simple matter to calculate the loss in wrongly rejecting a good batch. The difficulty arises in deciding the cost of wrongly accepting a bad batch since this is tied up with such difficult questions as the value to a manufacturer of a customer's goodwill. As we can note nowadays, in the wake of the COVID-19 pandemic, the manufacturer's goodwill with the customer comes into plan in reacting to supply chain disruptions.

OLIVER-BADIRU EXPERT SYSTEM FOR SUPPLIER DEVELOPMENT

Over three decades ago, Oliver and Badiru (1993) developed an expert system as a decision support system for supply chain management. Specifically, they developed an expert system for supplier development programs in manufacturing companies. The productive capacity of a manufacturer depends on the viability of its interfaces with its suppliers. A supplier development program (SDP) refers to a continuous process of identifying, assessing, classifying, supporting and developing suppliers with an objective to create and maintain a network of competent and consistent suppliers.

Organizational excellence would be better off if we could retain all the experts in various fields. Recognizing the limitations imposed by nature, Oliver and Badiru (1993) conducted research on knowledge acquisition, encoding, and dissemination. This implies the efficacy of expert systems. The Oliver-Badiru SDP software captures human expertise in the domain of supply chain. No manufacturing firm is totally self-sufficient. All firms depend on external suppliers to meet a variety of needs in the form of materials, goods, or services. On an average, manufacturers in North America spend over 60% of total revenue on acquisition of goods and supplies from outside suppliers. Factors like productivity, competitiveness, quality, and cost of products of a firm are affected by the performance of suppliers. With increasing global competition and the need for continuous improvement, the need for good suppliers is critical, particularly following an unfortunate disruption, such as what was caused by the COVID-19 pandemic in the year 2020. Having good suppliers is similar to having good business partners.

Oliver-Badiru SDP involves a variety of parameters, including quality, price, delivery, technology, geographical location, political environment, organizational policies, situational requirements, and other factors, attributes, and indicators, which further branch into numerous other considerations with differing levels of detail. Like other areas of materials management, it is a scenario of incompleteness, inconsistency, inaccuracy, uncertainty of data or relations, unstructured, complex and time-dependent procedures. The activities involved are hard or impossible to describe by quantitative tools, such as mathematical algorithms and repetitive processes. Managers in supplier development programs use a combination of facts, heuristics, and experience to tackle problems. All these are encoded in the Oliver-Badiru SDP software (Oliver & Badiru, 1993). In the 1990s, there was a lack of established methodologies for supplier development programs. The situation has vastly improved in recent years due to the emergence of more sophisticated computer hardware and software. Expert systems, which form a branch of artificial intelligence, have been the most productive and promising applications of AI in business and industry today (Badiru, 2021). The methodology of the SDP software has the following capabilities:

- Performs organizational assessment to detect if there is a need for SDP;
- Evaluates and rates suppliers based on the parameters defined and weights determined;
- Classifies suppliers into five groups based on ratings and user consultation;
- Offers advice on how to improve a supplier's performance.

Organizational assessment for SDP refers to analyzing the existing situation in the company and the suppliers, to detect if there is a need for SDP. A firm might need SDP due to one or more of the drivers listed below:

*   Establish a source for the new product or part;
*   Available supplier unwilling to supply;
*   Improve quality of incoming material;
*   Improve delivery performance;
*   Improve service;
*   Reduce cost of material;
*   Improve technical capability;
*   Reduce supplier base;
*   Enlarge supplier base;
*   Meet social, political, geographical, and environmental concerns;
*   Overcome market deficiency;
*   Change in the firm's operating systems (e.g., introduction of just in time [JIT]);
*   Large future requirement.

The organizational assessment module of the model has embedded rules which focus on meeting the objectives mentioned above. From user responses to questions about the above-mentioned objectives and comparison with the existing situation, the expert system makes recommendations. An example of a rule in SDP is presented below:

IF {Firm plans to introduce JIT: Just In Time}
THEN {SDP is recommended – Focus on JIT parameters during evaluation}

Using if-then-else rules within the expert system, the list of parameters embodied in SDP is summarized below:

Quality Improvement

1. Percentage of orders meeting design specification;
2. Percentage of in-process defects detected;
3. Percentage of downtime resulting from supplier error;
4. Quality of raw materials or parts used by supplier;
5. Supplier following world-class quality standards;
6. Participating in the firm's quality program.

Cost of Product

1. Percentage of difference between actual and budgeted price;
2. Price history (increase or decrease);
3. Number of cost reduction programs initiated by supplier;
4. Worldwide competitive cost structure;
5. Credit rating.

Delivery Commitments

1. Percentage of on-time deliveries;
2. Percentage of orders meeting quantity requirements;
3. Average time for normal orders;
4. Average time for special/rush orders;
5. Small delivery of items;
6. Frequent deliveries;
7. Ability to accommodate current volume.

Customer Service

1. Product and service availability;
2. Personnel capabilities;
3. Time required to correct supplier error;
4. Prompt and accurate reply to communications;
5. Advance written notice of any price change;
6. Advance notification about any delivery change.

Technical Capabilities

1. Manufacturing process;
2. Advanced facilities like CAD/CAM;
3. Cycle time reduction for new products and process;
4. Number of product improvements initiated by suppliers;
5. Engineering and product development support.

Operational Requirements

1. Availability of a common coding system;
2. Packaging, labeling, and shipping requirements;
3. Availability of proper paperwork;
4. Supplier ability to handle paper-based transactions;
5. Billing errors.

Strategic Alliance

1. Long-term co trunk meat for productivity and quality improvement;
2. Growth of the company;
3. Supplier financial at ability;
4. Ability to accommodate forecasted volume requirements;
5. Supplier willingness to dedicate capacity, resources;
6. Coordinate flow of data meeting objective of both organizations;
7. Cooperate with firm's supplier evaluation and rating system;
8. Respect confidentiality of orders;
9. Respect the terms and conditions of the contract;
10. Has electronic data interchange system;

11. Supplier ability to service JIT;
12. Early supplier development;
13. Suggest better or less expensive ways of using supplier product;
14. Considers the firm as a top customer;
15. Geographical proximity;
16. Social reasons like environment.

There are several parameters considered by managers in an SDP program to evaluate suppliers. These differ between organizations, products, and suppliers. Analysis and classification of these parameters involved is vital for developing a general model of SDP. The parameters involved are identified and classified during the development of SDP based on the prevailing methods available in the literature references of the 1990s. In the categorical method, for each parameter, the user assigns a (+), (−), or (0) for preferred, unsatisfactory, and neutral, respectively. This method is not effective, since all the parameters are weighted equally. The weighted point method assigns weights to the parameters, but all the performance measures act in uniform units, which is percentage. In reality, it is not possible to quantify many parameters to percentage. The cost-ratio method calculates cost ratios for several parameters, for different suppliers. It requires a comprehensive cost-accounting system. Dimensional analysis does numerical comparison of vendor performance with company standards and calculates an index based on weights assigned. But it is hard to convert many subjective data to numeric. The analytic hierarchy process (AHP) has also been used to model supplier evaluation (Oliver & Badiru, 1993). AHP requires pairwise comparison of all the suppliers for all the parameters. This is a tedious process when there is a large number of parameters involved. The evaluation technique used in SDP involves three steps:

1. Assigning weights to the parameters using AHP;
2. User-selected rating on a five-point scale (0–4) that best characterizes a supplier with aspect to each parameter;
3. Computing the final rating for each supplier.

Assigning weights: Initialization of weights to the parameters is done using AHP. AHP involves pairwise comparison of parameters with respect to their importance. Pairwise comparison is done separately for the classification titles and for the set of parameters under all classifications. The procedure involved in AHP is explained with the following example, computing the relative weights of the classification titles:

1. Strategic alliance
2. Quality improvement
3. Delivery commitments
4. Customer service
5. Technical capability
6. Cost of products
7. Operational requirements.

Pairwise comparisons of parameters are done for "importance to the firm." A matrix of pairwise comparisons (not shown here) is made based on user response to questions like the example presented below.

> *Question*: With respect to importance to your firm, how do you compare strategic alliance with quality improvement?
> *Options*: Absolutely less important (weight: 1/9); slightly less important (weight: 1/5); equally important (weight: 1); slightly more important (weight: 5); and absolutely more important (weight: 9).

Once the pairwise comparison is complete, normalization is done by dividing each entry in a column by the sum of all the entries in the column and the normalized average weights associated with each classification title are calculated. AHP is similarly done for all parameters, under all classifications, comparing with the other parameters in the classification. Thus, values for the following variables are obtained:

1. C_i – Weight of title of classification

2. P_{ij} – Weight of parameter

Where i = 1 to L (Number of classifications)
j = 1 to M (Number of parameters under a classification)

Rating suppliers: The rating module of the ES model presents the user with questions about all parameters. The user selects the best option among five which characterizes the supplier under evaluation. For example,

> *Question*: With respect to quality of raw materials or parts used by supplier, select the option which best characterizes the supplier.
> *Choices*: Poor (0), fair (1), good (2), very good (3), excellent (4).

The terminology used for choices in different questions for each parameter will differ, but the points assigned will remain the same, from 0 to 4 in all cases. This gives the provision to assign zero points to suppliers who do not qualify at all to score any points for a particular parameter. The rating process is repeated for all the parameters and for all the suppliers. Thus, values for the variables, Sijk, where k =1 to N (number of suppliers) are found. The final rating Rk of a supplier is computed using the formula shown below:

$$R_k = \sum_{i=1}^{L} \sum_{j=1}^{M} C_i P_{ij} S_{ijk}$$

Even though there are different numbers of parameters under each classification, since they are multiplied by the weights of the parameters, they are normalized and vary from 0 to 4. The maximum rating possible for a supplier is 4, and the minimum rating possible is 0. For an illustrative example of the SPD application to a practical supplier case example, please see Oliver and Badiru (1993).

SUPPLIER CLASSIFICATION

Based on the ratings obtained for each supplier, they are classified into five groups – (1) partner, (2) alliance, (3) annual renewal supplier, (4) normal supplier, and (5) unacceptable supplier. The required rating for each classification is decided by user.

SUPPLIER ADVISING

Supplier advising is the process of informing the suppliers about their classification, reasons for the classifications and suggestions on how they can improve. This module performs based on embedded rules which consist of advice for lack of specific parameters. A sample rule is as follows:

IF {Supplier has no ED I Facility}
AND {Firm's objective is to implement EDI}
THEN {Recommend: Get additional from plant manager

In a scenario of traditional buyer-supplier relationship being replaced with partner-partner relationship, SDP is vital for organizations to remain competitive. SDP involves many subjective but strategically pertinent data. Hence, an expert system model is useful for capturing and representing real-world scenarios. The parameters listed for organizational assessment and supplier evaluation are useful for organizations to identify specific parameters for their supplier development programs. This general Oliver-Badiru SDP model is adaptable for many situations and is suitable for modification to meet specific needs, particularly in the modern era of digital and global supply chain networks. Having a strong supplier development program can help retain, enhance, and prolong supplier relationships.

EFFICACY AND APPLICATIONS

Industrial engineering has a lot to offer in terms of tools and techniques applied to supply chain research, practice, and applications. The literature abounds with several examples. Loska and Higa (2020) address supply chain risk management for the future of organic supply chain for the US Air Force. Ekstrom et al. (2020) present differentiation strategies for defense supply chain design, with a specific focus on the Swedish defense system. Ellis et al. (2017) address the rising operational costs and software sustainment concerns in the US Air Force with respect to moving to newer technology for the Air Force standard base supply chain. With a base in the Netherlands, Van Strien et al. (2019) present a risk-management approach to performance-based contracting in military supply chains. Their paper investigates factors that influence service provider's willingness to accept risks induced by performance-based contracting. Food supply and food security, topics of particular focus for industrial engineers, are addressed by Lin et al. (2019) and Christensen et al. (2021). Overstreet et al. (2019) introduce a multi-study analysis of learning culture, human capital, and operational performance in supply chain management. The study empirically evaluates the relationship between learning culture, workforce level, human capital, and operational performance in two diverse supply chain populations, aircraft maintenance and logistics readiness. This study is particularly of interest in this book because of its learning curve alignment with the contents of Chapter 5. The contents addressing human capital and workforce level are also of special interest in relation to the earlier-era practice of POP (people off payroll) by ill-informed manufacturers. People are critical to the success of any supply chain. The COVID-19 pandemic lockdowns around the world make this fact painfully obvious. After the pandemic, employers found it difficult to rehire, retrain, and retain employees that were retrenched due to the pandemic. A lot of learning curve assets were consequently lost. Even for those who came back to work, the learn-forget-relearn processes (Badiru, 2015) came into play.

Within the realm of advanced research studies, Morrow (2021) and Femano (2021) present doctoral dissertation reports on the supply chain. According to Morrow (2021), information is crucial to supply chain performance because it is used to make decisions and trigger actions. Organizations across world-class supply chains increasingly use information technology to analyze and process supply chain data. However, supply chain management lacks a common language, making information exchange difficult. An ontology can provide a standardized framework that organizes a given knowledge domain. Morrow's research proposes a common language or ontology for supply chain management that can be understood by both humans and computers. This is an example of a human-systems integration research for supply chains. According to current research, an established and widely used supply chain framework is a good starting point for developing a supply chain ontology. Many researchers recommend using the Supply Chain Operations Reference (SCOR) model. This framework is translated into a software package that generates a Web Ontology Language (OWL), which can be used by information technology. Using SCOR 12.0 as the framework, an XML/OWL based model, Morrow developed a tool which can be used by information technology to improve information exchanges between supply chain partners.

In his own doctoral research, Femano (2021) emphasizes that businesses operate every day in a disruptive environment. Supply and demand uncertainty, natural disasters, global pandemics, and mishaps can all cause chaos to a supply chain's flow. It is impossible to predict every disruption a supply chain may encounter. The best an organization can do to protect network performance is to build resilience in the supply chain and lifeline of its operations. Ensuring that a supply chain has the proper built-in mechanisms to resist and recover from disruptions is referred to as supply chain resilience (SCR) (Femano, 2021). While it is generally agreed that SCR can be improved through the implementation of SCR strategies, the links between these strategies, performance improvement, and resilience is understudied. Femano's research focuses on resource-based view and theory of constraints to categorize the SCR strategies, examine the links between the strategies and performance, and develop a metric to measure network resilience over time. First, a meta-analytical study identifies generalizable relationships between SCR strategies and the organization's performance measures. Then, the SCR redundancy strategies are applied to a model simulation to illustrate the resilience curve response to different SCR strategic decisions. Resilience outcomes are compared using a developed resilience capability metric (RCM) utilizing area under the curve to measure the cumulative performance level of the system from disruption to a predetermined endpoint, representing how much of the system demand can be served by different network resilience designs. Finally, SCR flexibility strategies are analyzed to see how constraints imposed on a supply chain's response time could impact the resilience of the supply chain. Femano's work highlights the positive impact on performance and resilience that can be realized when organizations take the time to implement the proper SCR strategies, while providing managers with RCM to measure and compare the impact of different strategies within their organization.

Other relevant topics and ideas, pertinent for industrial engineering applications to the supply chain, are presented by Sinha et al. (2020), Perez (2013), Dolgui et al. (2018), Lin et al. (2019), and Hernández-Espallardo et al. (2010). A good illustration of the diverse application of industrial engineering to the supply chain of diverse industries is presented by Barve and Shanker (2020), in which they address green supply chain management. Their study focuses on detecting various parameters associated with green supply chain management practices in diamond mining industries globally.

Supply chain research, management, and application are within the operational spectrum of industrial engineering. Of particular relevance are the human-centric aspects of the practice of industrial engineering. The chapters that follow present a mix of quantitative and qualitative tools and techniques for supply chain analysis, management, and control.

REFERENCES

Badiru, A. B. (2015). Quality insights: Learning, forgetting, and relearning quality: A half-life learning curve modeling approach. *International Journal of Quality Engineering and Technology*, 5(1), 79–100.

Badiru, A. B. (2021). *Artificial intelligence and digital systems engineering*. Taylor & Francis/CRC Press.

Barve, A., & Shanker, S. (2020, March 10–12). *Sustainable supply chain concerns in diamond industries*. Proceedings of the International Conference on Industrial Engineering and Operations Management, Dubai, UAE.

Christensen, C., Wagner, T., & Langhals, B. (2021). Year-independent prediction of food insecurity using classical and neural network machine learning methods. *AI*, 2(1), 244–260. https://doi.org/10.3390/ai2020015

Dolgui, A., Ivanov, D., & Sokolov, B. (2018). Ripple effect in the supply chain: An analysis and recent literature. *International Journal of Production Research*, 56(1–2), 414–430.

Ekstrom, T., Hilletofth, P., & Skoglund, P. (2020). Differentiation strategies for defense supply chain design. *Journal of Defense Analytics and Logistics*, 4(2), 183–202. www.emerald.com/insight/2399-6439.htm

Ellis, T. L., Nicholson, R. A., Briggs, A. Y., Hunter, S. A., Harbison, J. E., Saladna, P. S., Garris, M. W., Ohnemus, R. K., O'Connor, J. E., & Reynolds, S. B. (2017). Lifting and shifting the air force retail supply system. *Journal of Defense Analytics and Logistics*, 1(2), 19–40. www.emerald.com/insight/2399-6439.htm

Femano, A. L. (2021). *Performance improvement through better understanding of supply chain resilience* [PhD dissertation, Air Force Institute of Technology].

Gilbreath, S. G., III. (1967). *A Bayesian procedure for the design of sequential sampling plans* [PhD dissertation, School of Industrial Engineering, Georgia Institute of Technology].

Hernández-Espallardo, M., Rodríguez-Orejuela, A., & Sánchez-Pérez, M. (2010). Inter-organizational governance, learning and performance in supply chains. *Supply Chain Management: An International Journal, 15*(2), 101–114. https://doi. org/10.1108/13598541011028714

Lin, X., Ruess, P. J., Marston, L., & Konar, M. (2019). Food flows between counties in the United States. *Environmental Research Letters, 14*(2), 2–18. https://doi.org/10.1088/1748-9326/ ab29ae

Loska, D., & Higa, J. (2020). The risk to reconstitution: Supply chain risk management for the future of the US air force's organic supply chain. *Journal of Defense Analytics and Logistics, 4*(1), 19–40. www.emerald.com/insight/2399-6439.htm

Morrow, D. (2021). *Developing a basic formal supply chain ontology to improve communication and interoperability* [PhD dissertation, Air Force Institute of Technology].

Oliver, G. E., & Badiru, A. B. (1993, November 18–19). *An expert system model for supplier development program in a manufacturing firm.* Proceedings of the 7th Oklahoma Symposium on Artificial Intelligence, Stillwater, Oklahoma, pp. 135–141.

Overstreet, R. E., Skipper, J. B., Huscroft, J. R., Cherry, M. J., & Cooper, A. L. (2019). Multstudy analysis of learning culture, human capital, and operational performance in supply chain management: The moderating role of workforce level. *Journal of Defense Analytics and Logistics, 3*(1), 41–59. www.emerald.com/insight/2399-6439.htm

Perez, H. D. (2013). *Supply chain roadmap: Aligning supply chain with business strategy* (D. H. Perez, Print.). www.SupplyChainRoadmap.com

Sinha, A., Bernardes, E., Calderon, R., & Wuest, T. (2020). *Digital supply networks.* McGraw-Hill.

Van Strien, J., Gelderman, C. J., & Semeijn, J. (2019). Performance-based contracting in military supply chains and the willingness to bear risks. *Journal of Defense Analytics and Logistics, 3*(1), 83–107. www.emerald.com/insight/2399-6439.htm

3 Flexibility and Resilience

INTRODUCTION TO MOVEMENTS IN THE GLOBAL SUPPLY CHAIN

The world is about moving items. A supply chain is the avenue through which those items are moved, physically or virtually, from one point to another. We often tend to think of a supply being about the movement of physical objects. However, in today's digitally leaning world, the movement of information digitally is as vital as moving physical objects. Therein lies the concept of a flexible supply chain, as advocated by this book. A flexible supply chain is the answer to operational disruptions in the movement of goods and services on a global scale. The global disruptions caused by wars and regional conflicts have shown how fragile and susceptible the global supply chain could be in the presence of a wide-sweeping disruption. Due to the pandemic, the normally smooth and efficient supply chain does not respond as swiftly and effectively as typically expected. The global supply chain contains interconnected pieces of supply routes and linkages. This creates a complex project scenario. Supply chain problems are not always due to product shortages, as consumers often think in their simplistic and uninformed ways. Below are some possible contributing factors, attributes, and indicators that may cause supply chain problems:

- Product shortages
- Transportation problems
- Workforce inequities
- Labor shortage
- Sabotage
- Accidents
- Natural disasters
- Political upheaval
- Hoarding
- Excessive demand vis-à-vis supply.

All of these are often recognized in their own individual rights to attention. But when they work in tandem to affect the supply chain, they can spell doom for national operations. The fluidity that may pose problems in the global supply chain can also turn out to present opportunities for alternate approaches that are responsive and adaptive to the current scenarios, operational needs, and market requirements. That is one justification for calling upon the tools and techniques of industrial engineering. Factory shutdowns in international sources, such as China, and shipping delays at domestic points create production inefficiencies that are best resolved through comprehensive industrial engineering techniques.

DOI: 10.1201/9781032620701-3   **51**

Because of the amorphous nature and fluidity of the supply chain, this book uses a triple-helix approach, leveraging industrial engineering, project management, and systems engineering. This calls for a mix of quantitative and qualitative techniques. So, the diverse soft and hard techniques favored by different readers are incorporated into the book.

From a comprehensive project management perspective, a supply chain that is designed to be robust in a steady-state scenario will be more likely to be resilient in a scenario of disruption. This is one of the key messages of this book. Using a project management framework and a systems approach, supply chain can be designed to be more robust, resilient, and responsive in both calm and chaotic scenarios.

INDUSTRIAL ENGINEERING EFFICACY IN THE SUPPLY CHAIN

Due to the multifaceted complexity of the global supply chain, the discipline of industrial engineering is well positioned to bring its tools and techniques to bear on the diverse challenges faced in the supply chain. Industrial engineering, a discipline that has flexibility and diversity of applicability, is defined as articulated below by what an industrial engineer does:

> An industrial engineer is one who is concerned with the design, installation, and improvement of integrated systems of people, materials, information, equipment, and energy by drawing upon specialized knowledge and skills in the mathematical, physical, and social sciences, together with the principles and methods of engineering analysis and design to specify, predict, and evaluate the results to be obtained from such systems.

This is exactly what the global supply chain needs in incorporating the multifaceted nuances of the interconnectedness of supply chains around the world.

BEYOND FOOD SUPPLY

Although food supply is often the most easily noticed disruption of the supply chain, there are many other supply chain problems, including the following:

- Auto parts
- Gasoline
- Computer chips
- Technical workforce.

In fact, this book initially started with a focus on the food supply chain. It was, subsequently, expanded to general supply chains due to the other realities that emerged as a result of the pandemic and other disruptive events. In most cases, the supply chain is characterized by the following elements:

- Prediction (to forecast market needs);
- Production (to meet the forecasted demand);
- Transportation (to get the goods out to the market);
- Distribution (to get the goods to the points of sale);

- Transaction (selling of the goods through wholesale, retail, etc.);
- Consumption (buyer's usage of the goods);
- Repetition (repeat of the entire process to keep the market going).

This cycle pertains not only to physical goods, but also to the provision of services, such as in healthcare, travel, security, emergency response, and recreation. In fact, the movement of people through the airport is a form of a supply chain process, and the cycle described above is directly applicable. Askin and Sefair (2021) present an analytical modeling technique that fits this supposition perfectly. Their work involves improving the efficiency of airport security screening checkpoints. The first step in matching capacity to demand is to predict demand. That is the forecasting of the number of passengers arriving in each time segment. While this may sound simple, the reality is more complicated due to data availability and inherent randomness. Any disruption in the whole transportation system involving a particular airport could easily upend the passenger flow at the local airport level. For a practical implementation of their approach, Askin and Sefair (2021) developed a quantitative model of the number of passengers arriving at security screening checkpoints for specific time intervals on various days. Other aspects of the methodology involve estimating queue lengths and wait times. This is a stand component of the application of operations research to problem scenarios. The work was done at the research Center for Accelerating Operational Efficiency (CAOE) at Arizona State University.

COMPLEXITY AND MALLEABILITY

The global supply chain is subject to complexity and malleability in case of external disruptions. One case example of the complicated global supply chain can be seen in the seafood industry. It is reported 80% of the seafood consumed in the United States is imported (Zomorodi & Geiran, 2021; NOAA, 2021). In the same breath, 25% of the seafood caught, raised, or farmed in the United States is sent overseas for processing before being imported back into the US market. This symbiotic but amorphous import-export relationship can create challenges even in steady-state times, not to talk of disrupted market times. With the increasing interest in healthier living, consumers are turning more and more to seafood. Thus, the global seafood demand and supply will continue to grow. Trade logistics and agreements may shift, but the basic uptick of supply and demand will continue. Global trade tensions, political uncertainties, pandemic-caused economic downturns, and social unrests may disrupt standard supply chain practices. However, new technological breakthroughs in aquaculture techniques and biosecurity may offer enhanced responses. A system view of the overall market landscape is essential for being responsive and adaptive to the changing demand and supply profiles.

INPUT-PROCESS-OUTPUT REPRESENTATION

Even for a simple market commodity, the supply chain is a complex chain of input, process, and output as depicted in the ICOM (input, constraints, outputs, and mechanisms) model shown in Figure 3.1.

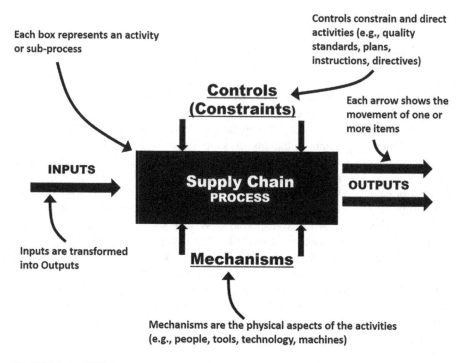

FIGURE 3.1   ICOM model for the supply chain.

INPUT

The supply chain starts with the market specifying the needs, requirements, conditions, and expectations. This is the input phase. Without knowing the nuances of the inputs, the supply chain will start out on the wrong foot. Even little things can matter bigly in a supply chain. This is akin to the age-old proverb by Benjamin Franklin:

> For the want of a nail the shoe was lost,
> For the want of a shoe the horse was lost,
> For the want of a horse the rider was lost,
> For the want of a rider the battle was lost,
> For the want of a battle the kingdom was lost,
> And all for the want of a horseshoe-nail.

From a systems perspective, all inputs of the market are important. It is in the input segment that the supply chain objectives and expected results are specified. Like the scope of work in project management, the scope of the supply chain market also is important to be specified as a part of the inputs.

CONSTRAINTS

There is no flawless or perfect system anywhere. Every supply chain will be subject to constraints, both internal and external. Constraints are the reality of any operation. Problems and impediments crop up and must be addressed promptly before disastrous consequences can develop. Some potential constraints include budget limitation, financial requirements, reporting expectations, schedule impositions, legal guidelines, contractual requirements, trade embargoes, environmental issues, quality standards, and government regulations. Using a systems approach, the linkages between the constraints can be better understood and managed. Supply chain objectives may have to modified in response to the prevailing constraints. Multi-constrained operation poses a challenge. But a mix of qualitative and quantitative tools can help resolve, overcome, or mitigate the challenges.

MECHANISM

In a supply chain, the mechanism is the people, knowledge base, tools, techniques, models, expertise, talents, technology, and capital. These constitute the "engine" required to move the supply chain forward. The list of mechanisms facilitating the forward movement of the supply chain will often be dynamic. The operators must be responsive and adaptive to adding new mechanisms. For example, the emergence of the COVID-19 pandemic necessitated adding new mechanisms (e.g., COVID testing and vaccination) to several supply chains.

OUTPUT

The output is like the "coin of the realm" of the supply chain enterprise. The output is the valued result of all the inner workings of the supply chain. Outputs can be defined in terms of the generation of a physical product, the provision of a service, or the achievement of a desirable result. Market surveys are important for accurately assessing market expectations and ascertaining the quality of the output. In implementation, the ICOM model provides guidance for specifying key result areas (KRA) and key performance indicators (KPI) which can guide the success and resilience of the supply chain.

The input-process-output framework of the ICOM model can be seen and felt at the gasoline pumps in local communities. Even when there is plenty of gasoline in the nation, getting it to the fuel pumps at gas stations is a different story. There are often reports of some stations running out of gasoline for short periods. The problem is often traced to a shortage of tanker truck drivers to transport the fuel from storage terminals to the fuel stations. From the consumer's perspective, the fuel needs to be where it is needed. But if the fuel is plentiful without truck drivers to drive the trucks to move the fuel to its respective points of sale, the fuel will be a "no-show" at the local level. Even shortages of short periods can create an annoyance, inconvenience, and predicament that can adversely affect human behavior, leading to events such as road rage. The whole thing is a complex system with several intervening layers of needs and consideration. This again affirms the need for a global systems view of the supply chain.

Apart from tanker trucks moving gasoline from point to point, there are other elements in the global supply chain that are not often physical seen at the local level. For example, during the COVID-19 pandemic, reports circulated about merchant ship crews stuck at sea or in ports. If there is no movement of international seafaring ships due to the pandemic, economic issues, trade wars, or other reasons, there will be no supply of goods to points in the global network. It is well known how computer chip shortages affect not only the computer production industry, but also the automobile industry and several feeder industries. We can all feel the pains of problems in the global supply chain.

PROJECT SYSTEMS FRAMEWORK FOR SUPPLY CHAIN

It is a systems world in the long run. Things work better with a systems approach. In the final analysis, every endeavor is a project. Applying an integration of the principles of systems management and project management provides a more robust handle of the overall supply chain. Thus, the rest of this chapter discusses various aspects of a project systems framework for the supply chain.

A systems view of a project makes the project execution more agile, efficient, and effective. A system is a collection of interrelated elements working together synergistically to achieve a set of objectives. Any project is a collection of interrelated activities, people, tools, resources, processes, and other assets brought together in the pursuit of a common goal. The goal may be in terms of generating a physical product, providing a service, or achieving a specific result. This makes it possible to view any project as a system that is amenable to all the classical and modern concepts of systems management.

Project management is the foundation of everything we do. Having knowledge is not enough, we must apply the knowledge strategically and systematically for it to be of any use. The knowledge must be applied to do something in the pursuit of objectives. Project management facilitates the application of knowledge and willingness to actually accomplish tasks. Where there is knowledge, willingness often follows. But it is the project execution that actually gets jobs accomplished. From the very basic tasks to the very complex endeavors, project management must be applied to get things done. It is, thus, essential that project management be a part of the core of every supply chain improvement initiative. The tools and techniques presented here are applicable to any project-oriented pursuit not only a supply chain, but also everywhere else. Everything that everyone does can be defined as a project. In this regard, a systems approach is of utmost importance in the global supply chain. Systems application encompasses the following elements:

- Technological systems (e.g., engineering control systems and mechanical systems);
- Organizational systems (e.g., work process design and operating structures);
- Human infrastructure (e.g., workforce development, interpersonal relationships, human-systems integration).

In systems-based project management, it is essential that related techniques be employed in an integrated fashion so as to maximize the total project output. One definition of systems project management offered here is stated as follows:

> Systems project management is the process of using systems approach to manage, allocate, and time resources to achieve systems-wide goals in an efficient and expeditious manner.

The definition calls for a systematic integration of technology, human resources, and work process design to achieve goals and objectives. There should be a balance in the synergistic integration of humans and technology. There should not be an overreliance on technology, nor should there be an overdependence on human processes. Similarly, there should not be too much emphasis on analytical models to the detriment of common-sense human-based decisions. Particularly in this era of digital operations, many cyber risks exist to upend even the best-designed systems. Recent hacking incidents around the world point to the need to exercise all human-based caution to mitigate digital-based risks.

Engineering problem-solving methodology can boost the readiness and resilience of supply chains. In this regard, the techniques of systems engineering have proven effective in addressing multifaceted challenges, even in non-engineering platforms. Systems engineering is growing in appeal as an avenue to achieve organizational goals and improve operational effectiveness and efficiency. Researchers and practitioners in business, industry, and government are collaboratively embracing systems engineering implementations. Several definitions of systems engineering exist. The following is one comprehensive definition:

> Systems engineering is the application of engineering to solutions of a multifaceted problem through a systematic collection and integration of parts of the problem with respect to the life cycle of the problem. It is the branch of engineering concerned with the development, implementation, and use of large or complex systems. It focuses on specific goals of a system considering the specifications, prevailing constraints, expected services, possible behaviors, and structure of the system. It also involves a consideration of the activities required to assure that the system's performance matches the stated goals. Systems engineering addresses the integration of tools, people, and processes required to achieve a cost-effective and timely operation of the system.

LOGISTICS IN PROJECT SYSTEMS

Logistics can be defined as the planning and implementation of a complex task, the planning and control of the flow of goods and materials through an organization or manufacturing process, or the planning and organization of the movement of personnel, equipment, and supplies. Complex projects represent a hierarchical system of operations. Thus, we can view a project system as collection of interrelated projects all serving a common end goal. Consequently, the following universal definition is applicable to supply chains:

> Project systems logistics is the planning, implementation, movement, scheduling, and control of people, equipment, goods, materials, and supplies across the interfacing boundaries of several related projects.

Conventional project management must be modified and expanded to address the unique requirements for logistics in project systems.

SYSTEM CONSTRAINTS

Any supply chain is a complex system with multiple constraints. The earlier the prevailing constraints are recognized and attended to the better for a smoother operation later on. Systems management is the pursuit of organizational goals within the constraints of time, cost, and quality expectations. The iron triangle, often referred to as the triple constraints, shows that project accomplishments are constrained by the boundaries of quality, time, and cost. In this case, quality represents the composite collection of project requirements. In a situation where precise optimization is not possible, there will have to be trade-offs between these three factors of success. The concept of the iron triangle is that a rigid triangle of constraints encases the project. Everything must be accomplished within the boundaries of time, cost, and quality. If a better quality is expected, a compromise along the axes of time and cost must be executed, thereby altering the shape or profile of the iron triangle.

The trade-off relationships are not linear and must be visualized in a multidimensional context. This is better articulated by a three-dimensional view of the systems constraints using applicable schematics of the supply chain characteristics. Scope requirements determine the project boundary and trade-offs must be done within that boundary.

INFLUENCE THEORY

Systems influence philosophy suggests the realization that you control the internal environment while only influencing the external environment. The internal (controllable) environment is represented as a black box in the typical input-process-output relationship. The outside (uncontrollable) environment is bounded by a cloud representation for the unknowns. In the comprehensive systems structure, inputs come from the global environment, are moderated by the immediate outside environment, and are delivered to the inside environment. In an unstructured internal environment, functions occur as blobs. A "blobby" environment is characterized by intractable activities where everyone is busy, but without a cohesive structure of input-output relationships. In such a case, the following disadvantages may be present:

- Lack of traceability
- Lack of process control
- Higher operating cost
- Inefficient personnel interfaces
- Unrealized technology potentials.

Organizations often inadvertently fall into the blobs structure because it is simple, low cost, and less time-consuming until a problem develops. A desired alternative is to model the project system using a systems value-stream structure, where each point-to-point interface is identifiable, traceable, and controllable, particularly if the project is implemented in choreographed increments. This suggests using a proactive and problem-preempting approach to execute projects. This alternative has the following advantages:

- Problem diagnosis is easier.
- Accountability is higher.
- Operating waste is minimized.
- Conflict resolution is faster.
- Value points are traceable.

MANAGEMENT BY SUPPLY

Every supply chain is a project and should be managed with the conventional tools and techniques of project management. The concept of management by supply opens the avenue to applying rigorous project management to supply chain challenges. Project management continues to grow as an effective means of managing functions of any form in any type of organization. Project management should be an enterprise-wide systems-based endeavor. Enterprise-wide project management is the application of project management techniques and practices across the full spectrum of the enterprise. This concept is also referred to as management by project (MBP). MBP is an effective approach that employs project management techniques in various functions within an organization. MBP recommends pursuing endeavors as project-oriented activities. In this respect every endeavor, large or small, simple or complex, can be modeled as a project and managed rigorously accordingly. Project management is an effective way to conduct any business activity. It represents a disciplined approach that defines any work assignment as a project. Under MBP, every undertaking is viewed as a project that must be managed just like a traditional project.

The traditional definition of a project as "a unique one-of-a-kind endeavor that has a definite beginning and a definite end" is still applicable in the sense that we can cobble together a series of such "definitely bounded" projects to achieve a composite project that spans multiple time periods and multiple end products. This was shown by Badiru (1988) in the novel application of project management to manufacturing, which was, hitherto, considered not to be a traditional project. The characteristics required of each project so defined are

1. An identified scope and a goal
2. A desired completion time
3. Availability of resources
4. A defined performance measure
5. A measurement scale for review of work.

An MBP approach to operations helps in identifying unique entities within functional requirements. This identification helps determine where functions overlap and how they are interrelated, thus paving the way for better planning, scheduling, and control. Enterprise-wide project management facilitates a unified view of organizational goals and provides a way for project teams to use information generated by other departments to carry out their functions.

The use of project management continues to grow rapidly. The need to develop effective management tools increases with the increasing complexity of new technologies and processes. The life cycle of a new product to be introduced into a competitive market is a good example of a complex process that must be managed with integrative project management approaches. The product will encounter management functions as it goes from one stage to the next. Project management will be needed throughout the design and production stages of the product. Project management will be needed in developing marketing, transportation, and delivery strategies for the product. When the product finally gets to the customer, project management will be needed to integrate its use with those of other products within the customer's organization.

The need for a project management approach is established by the fact that a project will always tend to increase in size even if its scope is narrowing. The following four literary laws are applicable to any project environment:

Parkinson's law: Work expands to fill the available time or space.
Peter's principle: People rise to the level of their incompetence.
Murphy's law: Whatever can go wrong will.
Badiru's rule: The grass is always greener where you most need it to be dead.

The 2020 pandemic confirmed that things can go wrong when we least expect. So, extra precautions are necessary. An integrated systems project management approach can help diminish the adverse impacts of these laws through good project planning, contingency readiness, organizing, scheduling, and control. Project management tools can be classified into three major categories:

1. *Qualitative tools*: These are the managerial tools that aid in the interpersonal and organizational processes required for project management.
2. *Quantitative tools*: These are analytical techniques that aid in the computational aspects of project management.
3. *Computer tools*: These are software and hardware tools that simplify the process of planning, organizing, scheduling, and controlling a project. Software tools can help in both the qualitative and quantitative analyses needed for project management.

Although individual books dealing with management principles, optimization models, and computer tools are available, there are few guidelines for the integration of the three areas for project management purposes. In this book, we integrate these three areas for a comprehensive guide to project management. The book introduces the *triad approach* to improve the effectiveness of project management with respect to schedule, cost, and performance constraints within the context of systems modeling. The approach considers not only the management of the project itself but also the management of all the functions that support the project. It is one thing to have a quantitative model, but it is a different thing to be able to apply the model to real-world problems in a practical form. The systems approach presented in this book illustrates how to make the transition from model to practice.

A systems approach helps increase the intersection of the three categories of project management tools and, hence, improve overall management effectiveness. Crisis should not be the instigator for the use of project management techniques. Project management approaches should be used upfront to prevent avoidable problems rather than to fight them when they develop. What is worth doing is worth doing well, right from the beginning.

SUCCESS FACTORS

The premise of this section is that the critical factors for systems success revolve around people and the personal commitment and dedication of each person. No matter how good a technology is and no matter how enhanced a process might be, it is ultimately the people involved that determine success. This makes it imperative to take care of people issues first in the overall systems approach to project management. Many organizations recognize this, but only few have been able to actualize the ideals of managing people productively. Execution of operational strategies requires forthrightness, openness, and commitment to get things done. Lip service and arm waving are not sufficient. Tangible programs that cater to the needs of people must be implemented. It is essential to provide incentives, encouragement, and empowerment for people to be self-actuating in determining how best to accomplish their job functions. A summary of critical factors for systems success encompasses the following:

- Total system management (hardware, software, and people)
- Operational effectiveness
- Operational efficiency
- System suitability
- System resilience
- System affordability
- System supportability
- System life cycle cost
- System performance
- System schedule
- System cost.

Systems engineering tools, techniques, and processes are essential for project life cycle management to make goals possible within the context of *SMART* principles, which are represented as follows:

1. *Specific*: Pursue specific and explicit outputs.
2. *Measurable*: Design of outputs that can be tracked, measured, and assessed.
3. *Achievable*: Make outputs to be achievable and aligned with organizational goals.
4. *Realistic*: Pursue only the goals that are realistic and result oriented.
5. *Timed*: Make outputs timed to facilitate accountability.

The general Project Management Body of Knowledge (PMBOK) was developed by the Project Management Institute (PMI). The body of knowledge comprises specific knowledge areas, which are organized into specific broad areas, all of which are applicable to a supply chain network.

1. Project *integration* management
2. Project *scope* management

3. Project *time* management
4. Project *cost* management
5. Project *quality* management
6. Project *human resource* management
7. Project *communications* management
8. Project *risk* management
9. Project *procurement and subcontract* management.

The listed segments of the body of knowledge of project management cover the range of functions associated with any project, particularly complex ones. Multinational projects particularly pose unique challenges pertaining to reliable power supply, efficient communication systems, credible government support, dependable procurement processes, consistent availability of technology, progressive industrial climate, trustworthy risk mitigation infrastructure, regular supply of skilled labor, uniform focus on quality of work, global consciousness, hassle-free bureaucratic processes, coherent safety and security system, steady law and order, unflinching focus on customer satisfaction, and fair labor relations. Assessing and resolving concerns about these issues in a step-by-step fashion will create a foundation of success for a large project. While no system can be perfect and satisfactory in all aspects, a tolerable trade-off on the factors is essential for project success.

The key components of each element of the body of knowledge are summarized as follows:

Integration Management

Integrative project charter
Project scope statement
Project management plan
Project execution management
Change control.

Scope Management

Focused scope statements
Cost/benefits analysis
Project constraints
Work breakdown structure
Responsibility breakdown structure
Change control.

Time Management

Schedule planning and control
Program evaluation and review technique (PERT) and Gantt charts
Critical path method
Network models
Resource loading
Reporting.

Cost Management

Financial analysis
Cost estimating
Forecasting
Cost control
Cost reporting.

Quality Management

Total quality management
Quality assurance
Quality control
Cost of quality
Quality conformance.

Human Resource Management

Leadership skill development
Team building
Motivation
Conflict management
Compensation
Organizational structures.

Communications Management

Communication matrix
Communication vehicles
Listening and presenting skills
Communication barriers and facilitators.

Risk Management

Risk identification
Risk analysis
Risk mitigation
Contingency planning.

Procurement and Subcontract Management

Material selection
Vendor prequalification
Contract types
Contract risk assessment
Contract negotiation
Contract change orders.

It should be noted that project life cycle is distinguished from product life cycle. Project life cycle does not explicitly address operational issues, whereas product life cycle is mostly about operational issues starting from the product's delivery to the end of its useful life. Note that for supply chain projects, the shape of the life cycle curve may be expedited due to the rapid developments that often occur in digital operations. For example, for a high technology project, the entire life cycle may be shortened, with a very rapid initial phase, even though the conceptualization stage may be very long. Typical characteristics of project life cycle include the following:

1. Cost and staffing requirements are lowest at the beginning of the project and ramp up during the initial and development stages.
2. The probability of successfully completing the project is lowest at the beginning and highest at the end. This is because many unknowns (risks and uncertainties) exist at the beginning of the project. As the project nears its end, there are fewer opportunities for risks and uncertainties.
3. The risks to the project organization (project owner) are lowest at the beginning and highest at the end. This is because not much investment has gone into the project at the beginning, whereas much has been committed by the end of the project. There is a higher sunk cost manifested at the end of the project.
4. The ability of the stakeholders to influence the final project outcome (cost, quality, and schedule) is highest at the beginning and gets progressively lower toward the end of the project. This is intuitive because influence is best exerted at the beginning of an endeavor.
5. Value of scope changes decreases over time during the project life cycle while the cost of scope changes increases over time. The suggestion is to decide and finalize scope as early as possible. If there are to be scope changes, do them as early as possible.

PROJECT STRUCTURE AND DESIGN

SUPPLY CHAIN PROJECT IDENTIFICATION

Project identification is the stage where a need for a proposed project is identified, defined, and justified. A project may be concerned with the development of new products, implementation of new processes, or improvement of existing facilities.

SUPPLY CHAIN PROJECT DEFINITION

Project definition is the phase at which the purpose of the project is clarified. A *mission statement* is the major output of this stage. For example, a prevailing low level of productivity may indicate a need for a new manufacturing technology. In general, the definition should specify how project management may be used to avoid missed deadlines, poor scheduling, inadequate resource allocation, lack of coordination, poor quality, and conflicting priorities.

SUPPLY CHAIN PROJECT PLANNING

A plan represents the outline of the series of actions needed to accomplish a goal. Project planning determines how to initiate a project and execute its objectives. It may be a simple statement of a project goal or it may be a detailed account of procedures to be followed during the project. Planning can be summarized as

Objectives
Project definition
Team organization
Performance criteria (time, cost, quality).

SUPPLY CHAIN PROJECT ORGANIZING

Project organization specifies how to integrate the functions of the personnel involved in a project. Organizing is usually done concurrently with project planning. Directing is an important aspect of project organization. Directing involves guiding and supervising the project personnel. It is a crucial aspect of the management function. Directing requires skillful managers who can interact with subordinates effectively through good communication and motivation techniques. A good project manager will facilitate project success by directing their staff, through proper task assignments, toward the project goal.

Workers perform better when there are clearly defined expectations. They need to know how their job functions contribute to the overall goals of the project. Workers should be given some flexibility for self-direction in performing their functions. Individual worker needs and limitations should be recognized by the manager when directing project functions. Directing a project requires skills dealing with motivating, supervising, and delegating.

SUPPLY CHAIN RESOURCE ALLOCATION

Project goals and objectives are accomplished by allocating resources to functional requirements. Resources can consist of money, people, equipment, tools, facilities, information, skills, and so on. These are usually in short supply. The people needed for a particular task may be committed to other ongoing projects. A crucial piece of equipment may be under the control of another team.

SUPPLY CHAIN PROJECT SCHEDULING

Timeliness is the essence of project management. Scheduling is often the major focus in project management. The main purpose of scheduling is to allocate resources so that the overall project objectives are achieved within a reasonable time span. Project objectives are generally conflicting in nature. For example, minimization of the project completion time and minimization of the project cost are conflicting objectives. That is, one objective is improved at the expense of worsening the other objective. Therefore, project scheduling is a multiple-objective decision-making problem.

In general, scheduling involves the assignment of time periods to specific tasks within the work schedule. Resource availability, time limitations, urgency level, required performance level, precedence requirements, work priorities, technical constraints, and other factors complicate the scheduling process. Thus, the assignment of a time slot to a task does not necessarily ensure that the task will be performed satisfactorily in accordance with the schedule. Consequently, careful control must be developed and maintained throughout the project scheduling process.

SUPPLY CHAIN PROJECT TRACKING

This phase involves checking whether or not project results conform to project plans and performance specifications. Tracking and reporting are prerequisites for project control. A properly organized report of the project status will help identify any deficiencies in the progress of the project and help pinpoint corrective actions.

PROJECT CONTROL

Project control requires that appropriate actions be taken to correct unacceptable deviations from expected performance. Control is actuated through measurement, evaluation, and corrective action. Measurement is the process of measuring the relationship between planned performance and actual performance with respect to project objectives. The variables to be measured, the measurement scales, and the measuring approaches should be clearly specified during the planning stage. Corrective actions may involve rescheduling, reallocation of resources, or expedition of task performance.

Tracking and reporting
Measurement and evaluation
Corrective action (plan revision, rescheduling, updating).

SUPPLY CHAIN PROJECT TERMINATION

Termination is the last stage of a project. The phaseout of a project is as important as its initiation. The termination of a project should be implemented expeditiously. A project should not be allowed to drag on after the expected completion time. A terminal activity should be defined for a project during the planning phase. An example of a terminal activity may be the submission of a final report, the power on of new equipment, or the signing of a release order. The conclusion of such an activity should be viewed as the completion of the project. Arrangements may be made for follow-up activities that may improve or extend the outcome of the project. These follow-up or spin-off projects should be managed as new projects but with proper input-output relationships within the sequence of projects.

PROJECT IMPLEMENTATION TEMPLATE

While this chapter advocates the main tenets of PMI's PMBOK, it also recommends the traditional project management framework encompassing the broad sequence summarized below:

Planning ➜ Organizing ➜ Scheduling ➜ Control ➜ Termination

An outline of the functions to be carried out during a project should be made during the planning stage of the project. A model for such an outline is presented hereafter. It may be necessary to rearrange the contents of the outline to fit the specific needs of a project.

Planning

1. Specify project background
   a. Define current situation and process
      i. Understand the process
      ii. Identify important variables
      iii. Quantify variables
   b. Identify areas for improvement
      i. List and discuss the areas
      ii. Study potential strategy for solution
2. Define unique terminologies relevant to the project
   a. Industry-specific terminologies
   b. Company-specific terminologies
   c. Project-specific terminologies
3. Define project goal and objectives
   a. Write mission statement
   b. Solicit inputs and ideas from personnel
4. Establish performance standards
   a. Schedule
   b. Performance
   c. Cost
5. Conduct formal project feasibility study
   a. Determine impact on cost
   b. Determine impact on organization
   c. Determine project deliverables
6. Secure management support.

Organizing

1. Identify project management team
   a. Specify project organization structure
      i. Matrix structure
      ii. Formal and informal structures
      iii. Justify structure
   b. Specify departments involved and key personnel
      i. Purchasing
      ii. Materials management
      iii. Engineering, design, manufacturing, and so on
   c. Define project management responsibilities
      i. Select project manager
      ii. Write project charter
      iii. Establish project policies and procedures

2. Implement Triple C model
   a. Communication
      i. Determine communication interfaces
      ii. Develop communication matrix
   b. Cooperation
      i. Outline cooperation requirements, policies, and procedures
   c. Coordination
      i. Develop work breakdown structure
      ii. Assign task responsibilities
      iii. Develop responsibility chart.

Scheduling (Resource Allocation)

1. Develop master schedule
   a. Estimate task duration
   b. Identify task precedence requirements
      i. Technical precedence
      ii. Resource-imposed precedence
      iii. Procedural precedence
   c. Use analytical models
      i. CPM
      ii. PERT
      iii. Gantt chart
      iv. Optimization models.

Control (Tracking, Reporting, and Correction)

1. Establish guidelines for tracking, reporting, and control
   a. Define data requirements
      i. Data categories
      ii. Data characterization
      iii. Measurement scales
   b. Develop data documentation
      i. Data update requirements
      ii. Data quality control
      iii. Establish data security measures
2. Categorize control points
   a. Schedule audit
      i. Activity network and Gantt charts
      ii. Milestones
      iii. Delivery schedule
   b. Performance audit
      i. Employee performance
      ii. Product quality
   c. Cost audit
      i. Cost containment measures
      ii. Percent completion versus budget depletion

3. Identify implementation process
   a. Comparison with targeted schedules
   b. Corrective course of action
      i. Rescheduling
      ii. Reallocation of resources.

Termination (Close, Phaseout)

1. Conduct performance review
2. Develop strategy for follow-up projects
3. Arrange for personnel retention, release, and reassignment.

Documentation

1. Document project outcome
2. Submit final report
3. Archive report for future reference.

LEAN OPERATIONS

Facing a lean period in supply chain project management creates value in terms of figuring out how to eliminate or reduce operational waste that is inherent in many human-governed processes. It is a natural fact that having to make do with limited resources creates opportunities for resourcefulness and innovation, which requires an integrated-systems view of what is available and what can be leveraged. The lean principles that are embraced by business, industry, and government have been around for a long time. It is just that we are now being forced to implement lean practices in diverse operational challenges due to the escalating shortage of resources. It is unrealistic to expect that problems that have rooted themselves in different parts of an organization can be solved by a single-point attack. Rather, a systematic probing of all the nooks and corners of the problem must be assessed and tackled in an integrated manner.

Contrary to the contention in some technocratic circles that budget cuts will stifle innovation, it is a fact that a reduction of resources often forces more creativity in identifying wastes and leveraging opportunities that lie fallow in nooks and crannies of an organization. This is not an issue of wanting more for less. Rather, it is an issue of doing more with less. It is through a systems viewpoint that new opportunities to innovate can be spotted. Necessity and adversity can, indeed, spark invention.

DECISION ANALYSIS

Systems decision analysis facilitates a proper consideration of the essential elements of decisions in a project systems environment. These essential elements include the problem statement, information, performance measure, decision model, and an implementation of the decision. The recommended steps are enumerated as follows:

STEP 1. PROBLEM STATEMENT

A problem involves choosing between competing, and probably conflicting, alternatives. The components of problem solving in project management include

Describing the problem (goals, performance measures);
Defining a model to represent the problem;
Solving the model;
Testing the solution;
Implementing and maintaining the solution.

Problem definition is very crucial. In many cases, *symptoms* of a problem are more readily recognized than its *cause* and *location*. Even after the problem is accurately identified and defined, a benefit/cost analysis may be needed to determine if the cost of solving the problem is justified.

STEP 2. DATA AND INFORMATION REQUIREMENTS

Information is the driving force for the project decision process. Information clarifies the relative states of past, present, and future events. The collection, storage, retrieval, organization, and processing of raw data are important components for generating information. Without data, there can be no information. Without good information, there cannot be a valid decision. The essential requirements for generating information are

Ensuring that an effective data collection procedure is followed;
Determining the type and the appropriate amount of data to collect;
Evaluating the data collected with respect to information potential;
Evaluating the cost of collecting the required data.

For example, suppose a manager is presented with a recorded fact that says, "Sales for the last quarter are 10,000 units." This constitutes ordinary data. There are many ways of using the aforementioned data to decide, depending on the manager's value system. An analyst, however, can ensure the proper use of the data by transforming it into information, such as "Sales of 10,000 units for the last quarter are within x percent of the targeted value." This type of information is more useful to the manager for decision making.

STEP 3. PERFORMANCE MEASURE

A performance measure for the competing alternatives should be specified. The decision maker assigns a perceived worth or value to the available alternatives. Setting measures of performance is crucial to the process of defining and selecting alternatives. Some performance measures, commonly used in project management are project cost, completion time, resource usage, and stability in the workforce.

STEP 4. DECISION MODEL

A decision model provides the basis for the analysis and synthesis of information, and is the mechanism by which competing alternatives are compared. To be effective, a decision model must be based on a systematic and logical framework for guiding project decisions. A decision model can be a verbal, graphical, or mathematical representation of the ideas in the decision-making process. A project decision model should have the following characteristics:

Simplified representation of the actual situation;
Explanation and prediction of the actual situation;
Validity and appropriateness;
Applicability to similar problems.

The formulation of a decision model involves three essential components:

Abstraction: Determining the relevant factors;
Construction: Combining the factors into a logical model;
Validation: Assuring that the model adequately represents the problem.

The basic types of decision models for project management are described next:

Descriptive models: These models are directed at describing a decision scenario and identifying the associated problem. For example, a project analyst might use a critical path method (CPM) network model to identify bottleneck tasks in a project.

Prescriptive models: These models furnish procedural guidelines for implementing actions. The triple C approach, for example, is a model that prescribes the procedures for achieving communication, cooperation, and coordination in a project environment.

Predictive models: These models are used to predict future events in a problem environment. They are typically based on historical data about the problem situation. For example, a regression model based on past data may be used to predict future productivity gains associated with expected levels of resource allocation. Simulation models can be used when uncertainties exist in the task durations or resource requirements.

Satisficing models: These are models that provide trade-off strategies for achieving a satisfactory solution to a problem, within given constraints. Goal programming and other multicriteria techniques provide good satisficing solutions. For example, these models are helpful in cases where time limitations, resource shortages, and performance requirements constrain the implementation of a project.

Optimization models: These models are designed to find the best available solution to a problem subject to a certain set of constraints. For example, a linear programming model can be used to determine the optimal product mix in a production environment.

In many situations, two or more of the aforementioned models may be involved in the solution of a problem. For example, a descriptive model might provide insights into the nature of the problem; an optimization model might provide the optimal set of actions to take in solving the problem; a satisficing model might temper the optimal solution with reality; a prescriptive model might suggest the procedures for implementing the selected solution; and a predictive model might forecast a future outcome of the problem scenario.

Step 5. Making the Decision

Using the available data, information, and the decision model, the decision maker will determine the real-world actions that are needed to solve the stated problem. A sensitivity analysis may be useful for determining what changes in parameter values might cause a change in the decision.

Step 6. Implementing the Decision

A decision represents the selection of an alternative that satisfies the objective stated in the problem statement. A good decision is useless until it is implemented. An important aspect of a decision is to specify how it is to be implemented. Selling the decision and the project to management requires a well-organized persuasive presentation. The way a decision is presented can directly influence whether or not it is adopted. The presentation of a decision should include at least the following: An executive summary, technical aspects of the decision, managerial aspects of the decision, resources required to implement the decision, cost of the decision, the time frame for implementing the decision, and the risks associated with the decision.

Systems decisions are often complex, diffuse, distributed, and poorly understood. No one person has all the information to make all decisions accurately. As a result, crucial decisions are made by a group of people. Some organizations use outside consultants with appropriate expertise to make recommendations for important decisions. Other organizations set up their own internal consulting groups without having to go outside the organization. Decisions can be made through linear responsibility, in which case one person makes the final decision based on inputs from other people. Decisions can also be made through shared responsibility, in which case, a group of people share the responsibility for making joint decisions. The major advantages of group decision making are listed as follows:

1. Facilitation of a systems view of the problem environment;
2. Ability to share experience, knowledge, and resources. Many heads are better than one. A group will possess greater collective ability to solve a given decision problem;
3. Increased credibility. Decisions made by a group of people often carry more weight in an organization;
4. Improved morale. Personnel morale can be positively influenced because many people have the opportunity to participate in the decision-making process;
5. Better rationalization. The opportunity to observe other people's views can lead to an improvement in an individual's reasoning process;
6. Ability to accumulate more knowledge and facts from diverse sources;
7. Access to broader perspectives spanning different problem scenarios;

8. Ability to generate and consider alternatives from different perspectives;
9. Possibility for a broader-based involvement, leading to a higher likelihood of support;
10. Possibility for group leverage for networking, communication, and political clout.

In spite of the much-desired advantages, group decision making does possess the risk of flaws. Some possible disadvantages of group decision making are listed as follows:

1. Difficulty in arriving at a decision;
2. Slow operating time frame;
3. Possibility for individuals' conflicting views and objectives;
4. Reluctance of some individuals in implementing the decision;
5. Potential for power struggle and conflicts among the group;
6. Loss of productive employee time;
7. Too much compromise may lead to less-than-optimal group output;
8. Risk of one individual dominating the group;
9. Overreliance on group process may impede agility of management to make decision fast;
10. Risk of dragging feet due to repeated and iterative group meetings.

BRAINSTORMING

Brainstorming is a way of generating many new ideas. In brainstorming, the decision group comes together to discuss alternate ways of solving a problem. The members of the brainstorming group may be from different departments, may have different backgrounds and training, and may not even know one another. The diversity of the participants helps create a stimulating environment for generating different ideas from different viewpoints. The technique encourages free outward expression of new ideas no matter how far-fetched the ideas might appear. No criticism of any new idea is permitted during the brainstorming session. A major concern in brainstorming is that extroverts may take control of the discussions. For this reason, an experienced and respected individual should manage the brainstorming discussions. The group leader establishes the procedure for proposing ideas, keeps the discussions in line with the group's mission, discourages disruptive statements, and encourages the participation of all members.

After the group runs out of ideas, open discussions are held to weed out the unsuitable ones. It is expected that even the rejected ideas may stimulate the generation of other ideas, which may eventually lead to other favored ideas. Guidelines for improving brainstorming sessions are presented as follows:

Focus on a specific decision problem.
Keep ideas relevant to the intended decision.
Be receptive to all new ideas.
Evaluate the ideas on a relative basis after exhausting new ideas.
Maintain an atmosphere conducive to cooperative discussions.
Maintain a record of the ideas generated.

DELPHI METHOD

The traditional approach to group decision making is to obtain the opinion of experienced participants through open discussions. An attempt is made to reach a consensus among the participants. However, open group discussions are often biased because of the influence of subtle intimidation from dominant individuals. Even when the threat of a dominant individual is not present, opinions may still be swayed by group pressure. This is called the "bandwagon effect" of group decision making.

The Delphi method attempts to overcome these difficulties by requiring individuals to present their opinions anonymously through an intermediary. The method differs from the other interactive group methods because it eliminates face-to-face confrontations. It was originally developed for forecasting applications, but it has been modified in various ways for application to different types of decision making. The method can be quite useful for project management decisions. It is particularly effective when decisions must be based on a broad set of factors. The Delphi method is normally implemented as follows:

1. *Problem definition*: A decision problem that is considered significant is identified and clearly described.
2. *Group selection*: An appropriate group of experts or experienced individuals is formed to address the particular decision problem. Both internal and external experts may be involved in the Delphi process. A leading individual is appointed to serve as the administrator of the decision process. The group may operate through the mail or gather together in a room. In either case, all opinions are expressed anonymously on paper. If the group meets in the same room, care should be taken to provide enough room so that each member does not have the feeling that someone may accidentally or deliberately observe their responses.
3. *Initial opinion poll*: The technique is initiated by describing the problem to be addressed in unambiguous terms. The group members are requested to submit a list of major areas of concern in their specialty areas as they relate to the decision problem.
4. *Questionnaire design and distribution*: Questionnaires are prepared to address the areas of concern related to the decision problem. The written responses to the questionnaires are collected and organized by the administrator. The administrator aggregates the responses in a statistical format. For example, the average, mode, and median of the responses may be computed. This analysis is distributed to the decision group. Each member can then see how their responses compare with the anonymous views of the other members.
5. *Iterative balloting*: Additional questionnaires based on the previous responses are passed to the members. The members submit their responses again. They may choose to alter or not to alter their previous responses.
6. *Silent discussions and consensus*: The iterative balloting may involve anonymous written discussions of why some responses are correct or incorrect. The process is continued until a consensus is reached. A consensus may be

declared after five or six iterations of the balloting or when a specified percentage (e.g., 80%) of the group agrees on the questionnaires. If a consensus cannot be declared on a particular point, it may be displayed to the whole group with a note that it does not represent a consensus.

In addition to its use in technological forecasting, the Delphi method has been widely used in other general decision making. Its major characteristics of anonymity of responses, statistical summary of responses, and controlled procedure make it a reliable mechanism for obtaining numeric data from subjective opinion. The major limitations of the Delphi method are

1. Its effectiveness may be limited in cultures where strict hierarchy, seniority, and age influence decision-making processes.
2. Some experts may not readily accept the contribution of nonexperts to the group decision-making process.
3. Since opinions are expressed anonymously, some members may take the liberty of making ludicrous statements. However, if the group composition is carefully reviewed, this problem may be avoided.

NOMINAL GROUP TECHNIQUE

The nominal group technique is a silent version of brainstorming. It is a method of reaching consensus. Rather than asking people to state their ideas aloud, the team leader asks each member to jot down a minimum number of ideas, for example, five or six. A single list of ideas is then written on a chalkboard for the whole group to see. The group then discusses the ideas and weeds out some iteratively until a final decision is made. The nominal group technique is easier to control. Unlike brainstorming, where members may get into shouting matches, the nominal group technique permits members to silently present their views. In addition, it allows introversive members to contribute to the decision without the pressure of having to speak out too often.

In all of the group decision-making techniques, an important aspect that can enhance and expedite the decision-making process is to require that members review all pertinent data before coming to the group meeting. This will ensure that the decision process is not impeded by trivial preliminary discussions. Some disadvantages of group decision making are as follows:

1. Peer pressure in a group situation may influence a member's opinion or discussions.
2. In a large group, some members may not get to participate effectively in the discussions.
3. A member's relative reputation in the group may influence how well their opinion is rated.
4. A member with a dominant personality may overwhelm the other members in the discussions.

5. The limited time available to the group may create a time pressure that forces some members to present their opinions without fully evaluating the ramifications of the available data.
6. It is often difficult to get all members of a decision group together at the same time.

Despite the noted disadvantages, group decision making definitely has many advantages that may nullify the shortcomings. The advantages as presented earlier will have varying levels of effect from one organization to another. Team work can be enhanced in group decision making by adhering to the following guidelines:

1. Get a willing group of people together.
2. Set an achievable goal for the group.
3. Determine the limitations of the group.
4. Develop a set of guiding rules for the group.
5. Create an atmosphere conducive to group synergism.
6. Identify the questions to be addressed in advance.
7. Plan to address only one topic per meeting.

For major decisions and long-term group activities, arrange for team training that allows the group to learn the decision rules and responsibilities together. The steps for the nominal group technique are as follows:

1. Silently generate ideas, in writing.
2. Record ideas without discussion.
3. Conduct group discussion for clarification of meaning, not argument.
4. Vote to establish the priority or rank of each item.
5. Discuss vote.
6. Cast final vote.

INTERVIEWS, SURVEYS, AND QUESTIONNAIRES

Interviews, surveys, and questionnaires are important information gathering techniques. They also foster cooperative working relationships. They encourage direct participation and inputs into project decision-making processes. They provide an opportunity for employees at the lower levels of an organization to contribute ideas and inputs for decision making. The greater the number of people involved in the interviews, surveys, and questionnaires, the more valid the final decision. The following guidelines are useful for conducting interviews, surveys, and questionnaires to collect data and information for project decisions:

1. Collect and organize background information and supporting documents on the items to be covered by the interview, survey, or questionnaire.
2. Outline the items to be covered and list the major questions to be asked.
3. Use a suitable medium of interaction and communication: Telephone, fax, electronic mail, face-to-face, observation, meeting venue, poster, or memo.

4. Tell the respondent the purpose of the interview, survey, or questionnaire, and indicate how long it will take.
5. Use open-ended questions that stimulate ideas from the respondents.
6. Minimize the use of yes-or-no type of questions.
7. Encourage expressive statements that indicate the respondent's views.
8. Use the who, what, where, when, why, and how approach to elicit specific information.
9. Thank the respondents for their participation.
10. Let the respondents know the outcome of the exercise.

MULTIVOTE

Multivoting is a series of votes used to arrive at a group decision. It can be used to assign priorities to a list of items. It can be used at team meetings after a brain-storming session has generated a long list of items. Multivoting helps reduce such long lists to a few items, usually three to five. The steps for multivoting are as follows.

1. Take a first vote. Each person votes as many times as desired, but only once per item.
2. Circle the items receiving a relatively higher number of votes (i.e., majority vote) than the other items.
3. Take a second vote. Each person votes for a number of items equal to one-half the total number of items circled in step 2. Only one vote per item is permitted.
4. Repeat steps 2 and 3 until the list is reduced to three to five items depending on the needs of the group. It is not recommended to multivote down to only one item.
5. Perform further analysis of the items selected in step 4, if needed.

In terms of a summary of a project formulation of the supply chain, systems integration is the synergistic linking together of the various components, elements, and subsystems of a system, where the system may be a complex project, a large endeavor, or an expansive organization. Activities that are resident within the system must be managed both from the technical and managerial standpoints. Any weak link in the system, no matter how small, can be the reason that the overall system fails. In this regard, every component of a project is a critical element that must be nurtured and controlled. Embracing the systems principles for project management will increase the likelihood of success of projects.

REFERENCES

Askin, R. G., & Sefair, J. A. (2021, May). Improving the efficiency of airport security screening checkpoints. *ISE Magazine, 53*(5), 26–31.
Badiru, A. B. (1988). *Project management in manufacturing and high technology operations.* John Wiley & Sons.

NOAA – National Oceanic and Atmospheric Administration. (2021). *Fishwatch: U.S. seafood facts*. Retrieved June 26, 2021, from www.fishwatch.gov/sustainable-seafood/the-global-picture

Zomorodi, M., & Geiran, F. (2021). *Ayana Elizabeth Johnson: What should you look for when shopping for seafood?* [NPR Radio Hour podcast]. Retrieved June 26, 2021, from www.wvxu.org/post/ayana-elizabeth-johnson-what-should-you-look-when-shopping-seafood#stream/0

4 Supply Chain Systems Integration

INTRODUCTION TO THE DEJI SYSTEMS MODEL

What appears great in design may prove to be a failure in operation. This is why a Design concept must be followed by Evaluation, Justification, and Integration stages. Steve Jobs cautioned us about this system's reality:

> Design is not how it looks and feels, it's how it works:

Each supply chain is composed of a large array of elements, some simple and discernible and some are complex and subtle. Integrating the elements structurally provides for a flexible system. This chapter introduces the DEJI Systems Model, which is applicable for the Design, Evaluation, Justification, and Integration of supply chain elements. The overriding theme of the DEJI Systems Model is to recognize and respect the respective roles of people, policy, and technology in a flexible supply chain. Figure 4.1 illustrates how these elements interface and interact.

Ideas (designs, concepts, proposals, etc.) that are good in principle often don't integrate well into the realities of a sustainable implementation. Therein lies the case for systems integration as advocated by the DEJI Systems Model. As an author, my love of systems integration is predicated on the belief that you can fight to get what you want and get hurt in the fighting process, or you can systematically negotiate and win without shedding any tears, limited blood, or excessive sweat. It is in systems integration that we actually get things done. In a systems context, integration is everything. Several practical examples exist to ginger our interest in systems integration. One common (and relatable) example is in traffic behavior of drivers. Integration implies monitoring, adapting to, and reacting to the prevailing scenario in interstate road traffic. In this case, integration calls for observing and enmeshing into the ongoing traffic with respect to flow and speed. If everyone drives with a sense of traffic systems integration, there will be fewer accidents on the road. In the full DEJI systems framework, a driver would observe and design (i.e., formulate) driving actions, evaluate the rationale and potential consequences of the "designed" actions, justify (mentally) why the said actions are necessary and pertinent for the current traffic scenario, and then integrate the actions to the driving condition.

Integration is the basis for the success of any system. This is particularly critical for system of systems (SoS) applications where there may be many moving parts. This book presents a practical framework for applying the trademarked DEJI Systems Model for SoS, with respect to systems Design, Evaluation, Justification, and Integration. Case examples will be presented on recent applications of the model and its efficacy diverse problem scenarios. The DEJI Systems Model has been reported to

DOI: 10.1201/9781032620701-4

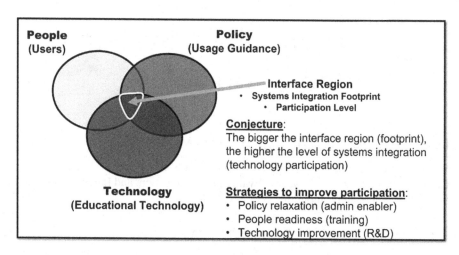

FIGURE 4.1   Interfacing roles of people, policy, and technology in a flexible supply chain.

be particularly effective for systems thinking in project planning, project evaluation, and project control and other human pursuits.

A familiar example is the January 2022 announcement of the retirement of US Supreme Court Justice Stephen Breyer. Justice Breyer was known as a legal pragmatist, who worked to make the law work in consonance with how the society live and work. That is a good spirit of aligning and integrating the law with people's lives and expectations. When a system is aligned with reality, the probability of success increases exponentially.

Badiru (2012) first formally introduced the trademarked DEJI Systems Model as a structured process for accomplishing systems Design, Evaluation, Justification, and Integration in product development. The model has since been adopted and applied to other application areas, such as quality management (Badiru, 2014) and engineering curriculum integration (Badiru & Racz, 2018). The premise of the model is that integration across a system is the overriding requirement for a successful system of systems (SoS).

A system is represented as consisting of multiple parts, all working together for a common purpose or goal. Systems can be small or large, simple or complex. Small devices can also be considered systems. Systems have inputs, processes, and outputs. Systems are usually explained using a model for a visual clarification inputs, process, and outputs. A model helps illustrate the major elements and their relationships. Figure 4.2 illustrates the basic model structure of a system.

In the ICOM chart, processes are the intervening mechanisms that convert inputs to outputs. This implies that processes represent the interactions that occur between elements in a system. In the human terms in the system, these may be viewed as the human facilitators and mediators in the system. When interrelated structures of input-process-output systems are stringed together, we have a system of systems that must be integrated structurally. Therein lies the efficacy of the integration component of the DEJI Systems Model.

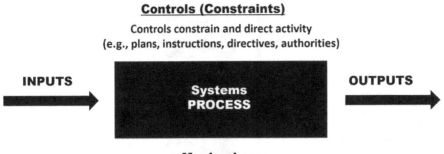

Controls (Constraints)

Controls constrain and direct activity
(e.g., plans, instructions, directives, authorities)

INPUTS

**Systems
PROCESS**

OUTPUTS

Mechanisms

Mechanisms are the physical aspects of the activities
(e.g., people, resources, space, budget, etc.)

FIGURE 4.2 ICOM systems input, process, and output structure.

Systems engineering is the application of engineering tools and techniques to the solutions of multifaceted problems through a systematic collection and integration of parts of the problem with respect to the life cycle of the problem. It is the branch of engineering concerned with the development, implementation, and use of large or complex data sets across diverse domains. It focuses on specific goals of a system considering the specifications, prevailing constraints, expected services, possible behaviors, and structure of the system. It also involves a consideration of the activities required to ensure that the system's performance matches specified goals. Systems engineering addresses the integration of tools, people, and processes required to achieve a cost-effective and timely operation of the system. Some of the features of this book include solutions to multifaceted problems; a holistic view of a problem domain; applications to both small and large problems; decomposition of complex problems into smaller, manageable chunks; direct considerations for the pertinent constraints that exist in the problem domain systematic linking of inputs to goals and outputs; explicit treatment of the integration of tools, people, and processes; and a compilation of existing systems engineering models. A typical decision support model is a representation of a system, which can be used to answer questions about the system. While systems engineering models facilitate decisions, they are not typically the conventional decision support systems. The end result of using a systems engineering approach is to integrate a solution into the normal organizational process. For that reason, the DEJI Systems Model is desired for its structured framework of Design, Evaluation, Justification, and Integration. The flowchart for this framework is shown in Figure 4.3.

SYSTEMS ENGINEERING COMPETENCY FRAMEWORK

The DEJI Systems Model is operationally aligned with the ideals and expectations of the INCOSE Systems Engineering Competency Framework (INCOSE SECF) developed by the International Council on Systems Engineering (INCOSE). INCOSE SECF represents a worldview of five competency groupings with 36 competencies

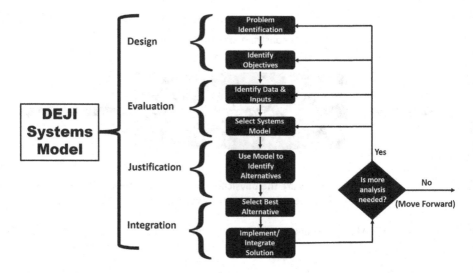

FIGURE 4.3   Flowchart of DEJI Systems Model flow framework.

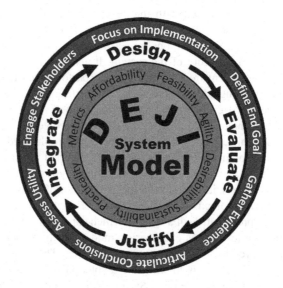

FIGURE 4.4   Trademarked logo of DEJI Systems Model.

for systems implementations. The integration requirement of the DEJI Systems Model directly focuses on what is required to achieve successful systems implementations. Figure 4.4 illustrates the wheel of the alignment of the DEJI Systems Model to the scope of practical systems implementations in business, industry, and other enterprises.

SYSTEM DESIGN

In as much as "design" is the origin of any system, the DEJI Systems Model starts with a structural articulation of what design embraces. In general, design can encompass a variety of pursuits ranging from physical product design to the design of conceptual framework for process improvement. In each generic frame, the systems value model (SVM) approach is effective in assessing the properties, expected characteristics, and desired qualities of a system. SVM provides an analytical decision aid for comparing system alternatives. In this case, value is represented as a p-dimensional vector:

$$V = f\left(A_1, A_2, \ldots, A_p\right),$$

Where $A = \left(A_1, \ldots, A_n\right)$ is a vector of quantitative measures of tangible and intangible attributes of the system. Examples of process attributes are quality, throughput, capability, productivity, cost, and schedule. Attributes are considered to be a combined function of factors, x_1, expressed as:

$$A_k\left(x_1, x_2, \ldots, x_{m_k}\right) = \sum_{i=1}^{m_k} f_i\left(x_i\right)$$

where $\{x_i\}$ = set of m factors associated with attribute A_k $(k = 1, 2, \ldots, p)$ and f_i = contribution function of factor x_i to attribute A_k. Examples of factors include reliability, flexibility, user acceptance, capacity utilization, safety, and design functionality. In fact, factors can be expanded to cover many of the common "ilities" of a system, such as affordability, practicality, desirability, configurability, modularity, desirability, maintainability, testability, reachability, and agility. Factors are themselves considered to be composed of indicators, v_i, expressed as

$$x_i\left(v_1, v_2, \ldots, v_n\right) = \sum_{j=1}^{n} z_i\left(v_i\right)$$

where $\{v_j\}$ = set of n indicators associated with factor x_i $(i = 1, 2, \ldots, m)$ and z_j = scaling function for each indicator variable v_j. Examples of indicators are project responsiveness, lead time, learning curve, and work rejects. By combining the above definitions, a composite measure of the value of a process can be modeled as:

$$V = f\left(A_1, A_2, \ldots, A_p\right)$$
$$= f\left\{\left[\sum_{i=1}^{m_1} f_i\left(\sum_{j=1}^{n} z_j\left(v_j\right)\right)\right]_1, \left[\sum_{i=1}^{m_2} f_i\left(\sum_{j=1}^{n} z_j\left(v_j\right)\right)\right]_2, \ldots, \left[\sum_{i=1}^{m_k} f_i\left(\sum_{j=1}^{n} z_j\left(v_j\right)\right)\right]_p\right\}$$

where m and n may assume different values for each attribute. A subjective measure to indicate the utility of the decision maker may be included in the model by using an attribute weighting factor, w_i, to obtain a weighted PV:

$$PV_w = f\left(w_1 A_1, w_2 A_2, \ldots, w_p A_p\right),$$

Where

$$\sum_{k=1}^{p} w_k = 1, \qquad \left(0 \le w_k \le 1\right).$$

With this modeling approach, a set of design options can be compared on the basis of a set of attributes and factors, both quantitative and qualitative. To illustrate the model above, suppose three IT options are to be evaluated based on four attribute elements: *Capability, suitability, performance*, and *productivity*. For this example, based on the equations, the value vector is defined as:

$$V = f\left(capability, suitability, performance, productivity\right)$$

Capability: The term "capability" refers to the ability of IT equipment to satisfy multiple requirements. For example, a certain piece of IT equipment may only provide computational service. A different piece of equipment may be capable of generating reports in addition to computational analysis, thus increasing the service variety that can be obtained. In the analysis, the levels of increase in service variety from the three competing equipment types are 38%, 40%, and 33%, respectively.

Suitability: "Suitability" refers to the appropriateness of the IT equipment for current operations. For example, the respective percentages of operating scope for which the three options are suitable for are 12%, 30%, and 53%.

Performance: "Performance," in this context, refers to the ability of the IT equipment to satisfy schedule and cost requirements. In the example, the three options can, respectively, satisfy requirements on 18%, 28%, and 52% of the typical set of jobs.

Productivity: "Productivity" can be measured by an assessment of the performance of the proposed IT equipment to meet workload requirements in relation to the existing equipment. For the example, the three options, respectively, show normalized increases of 0.02, −1.0, and −1.1 on a uniform scale of productivity measurement. Option C is the best "value" alternative in terms of suitability and performance. Option B shows the best capability measure, but its productivity is too low to justify the needed investment. Option A offers the best productivity, but its suitability measure is low. The analytical process can incorporate a lower control limit into the quantitative assessment such that any option providing value below that point will not be acceptable. Similarly, a minimum value target can be incorporated into the graphical plot such that each option is expected to exceed the target point on the value scale. The relative weights used in many justification methodologies are based on subjective propositions of decision makers. Some of those subjective weights can be enhanced by the incorporation of utility models. For example, the weights could be obtained from utility functions. There is a risk of spending too much time maximizing inputs at "point-of-sale" levels with little time defining and refining outputs at the "wholesale" systems level.

A systems view of operations is essential for every organization. Without a systems view, we cannot be sure we are pursuing the right outputs that can be *integrated* into the prevailing operating environment. Thus, the DEJI Systems Model allows for a multi-dimensional analysis of any endeavor, considering many of the typical "ilities" related to system of systems.

SYSTEM EVALUATION

Evaluation is the second stage of the structural application of the DEJI Systems Model. The evaluation of a system pursuit can range over a variety of rubrics related to organizational performance metrics. In many cases the basic requirement of as economic evaluation may be needed beyond the technical sphere provided in the design phase of the DEJI Systems Model. Other basis of the evaluation of a system may involve materials availability, quality expectation, supply chain reliability, skilled workforce, and so on. For example, an evaluation of a skilled workforce to run a well-designed system may be needed to ensure that the system can, indeed, be operated. In this case, operations research techniques of optimizing work assignment could be useful. For example, a resource-assignment algorithm can be used to enhance the quality of resource-allocation decisions. Suppose there are n tasks which must be performed by n workers. The cost of worker i performing task j is c_{ij}. It is desired to assign workers to the tasks in a fashion that minimizes the cost of completing the tasks. This problem scenario is referred to as the assignment problem. The technique for finding the optimal solution to the problem is called the assignment method. The assignment method is an iterative procedure that arrives at the optimal solution by improving on a trial solution at each stage of the procedure. Although the assignment method is cost-based, task duration can be incorporated into the modeling in terms of time-cost relationships for a more robust evaluation, based on the organization's interest. If the objective is to minimize the cost of the system implementation, the formulation of the assignment problem can be as shown below:

Let

$x_{ij} = 1$ if worker i is assigned to task j, $j = 1, 2, \ldots, n$,

$x_{ij} = 0$ if worker i is not assigned to task j

$c_{ij} = $ cost of worker i performing task j.

$$\text{Minimize:} \quad z = \sum_{i=1}^{n} \sum_{j=1}^{n} c_{ij} x_{ij}$$

$$\text{Subject to:} \quad \sum_{j=1}^{n} x_{ij} = 1, \quad i = 1, 2, \ldots, n$$

$$\sum_{i=1}^{n} x_{ij} = 1, \quad j = 1, 2, \ldots, n$$

$$x_{ij} \geq 0, \quad i, j = 1, 2, \ldots, n$$

The above formulation uses the non-negativity constraint, $x_{ij} \geq 0$, instead of the integer constraint, $x_{ij} = 0$ or 1. However, the solution of the model will still be integer-valued. Hence, the assignment problem is a special case of the common transportation problem in operations research, with the number of sources (m) = number of targets (n), $S_i = 1$ (supplies), and $D_i = 1$ (demands). The basic requirements of an assignment problem are summarized below:

1. There must be two or more tasks to be completed.
2. There must be two or more resources that can be assigned to the tasks.
3. The cost of using any of the resources to perform any of the tasks must be known.
4. Each resource is to be assigned to one and only one task.

If the number of tasks to be performed is greater than the number of workers available, we will need to add *dummy workers* to balance the problem. Similarly, if the number of workers is greater than the number of tasks, we will need to add *dummy tasks* to balance the problem. If there is no problem of overlapping, a worker's time may be split into segments so that a worker can be assigned to more than one task. In this case, each segment of the worker's time will be modeled as a separate resource in the assignment problem. Thus, the assignment problem can be extended to consider partial allocation of resource units to multiple tasks. The example presented here is just to illustrate what the evaluation stage of the DEJI Systems Model may entail. Other evaluation parameters and quantitative solution approaches are available for users to consider, based on internal interests and tools available within the organization.

SYSTEM JUSTIFICATION

Justification is the third stage of the application of the DEJI Systems Model. In this case, justification goes beyond the typical economic justification in project planning and control. While economic feasibility may be included in the evaluation stage of DEJI Systems Model, it is not the necessary basis of a systems justification. Justification, in the context of the DEJI System Model, is often more qualitative and conceptual than quantitative. Not all systems that are well designed and favorably evaluated are justified for implementation. Questions related to systems justification may include the following:

- Desirability of the system for the operating environment;
- Acceptability of the system by those who will be charged to run the system;
- Safety protocols related to operating the system;
- Regulatory oversight for operating the system;
- Compliance with industry standards for operating the system;
- Organizational philosophy about system expansion;
- Investment potential for actualizing the system;
- Sustainability potential for the new system.

The beauty of the DEJI Systems Model is that it explicitly requires or "forces" an organization to justify why a new is required. In many practical applications of systems modeling, some discoveries and revelations have pointed to discouraging realities that were not previously recognized or highlighted during the system realization stage (Valerdi, 2014).

SYSTEMS INTEGRATION

Integration is the fourth and last stage of the application of the DEJI Systems Model, and deferentially the most critical part of any systems implementation. A system that is not appropriately anchored to the prevailing operating environment may be doomed to failure. This fact is often the reason behind many system failures that are seen in practice. The DEJI System Model explicitly requires that any system of interest be *integrated* into its point and manner of use. Integration can be affirmed through qualitative or quantitative mechanisms. Some of the pertinent questions for integration include:

- Is the new system aligned with the existing operating framework of the organization?
- Is the new system in line with the prevailing mission of the organization?
- Can this system really do what it is designed to do?
- Is the prevailing work environment able to assimilate an implementation of the new system?
- What are the practical implementation requirements for adopting the new system?
- What operating constraints exist to successfully implement the new system?

With probing questions such as the above, DEJI System Model can positively impact how new systems are brought online in the prevailing operating environment, considering the existing tools, techniques, processes, and workforce with the organization. Without being integrated, a system will be in isolation and it may be worthless. We must integrate all the elements of a system on the basis of alignment of functional goals. In the case of a quantitative process, the overlap of systems for integration purposes can be viewed, conceptually, as projection integrals by considering areas bounded by the common elements of sub-systems. This approach is useful for system of systems applications, where each independent system is a sub-system of the larger SoS. Quantitative metrics can be applied at this stage for effective assessment of the composite system. Trade-off analysis is essential in system integration. Pertinent questions include the following:

- What level of trade-offs on the level of integration is tolerable?
- What is the incremental cost of pursuing higher integration?
- What is the marginal value of higher integration?
- What is the adverse impact of a failed integration attempt?

What is the value of integration of system characteristics over time? In this respect, an integral of the form below may be suitable for a mathematical exposition:

$$I = \int_{t_1}^{t_2} f(q)dq,$$

Where I = integrated value of quality, $f(q)$ = functional definition of quality, t_1 = initial time, and t_2 = final time within the planning horizon. An illustrative example is provided in Figure 4.5 for the case of a geometric alignment of the hypothetical physical parts of SoS.

Presented below are guidelines and important questions relevant for system integration.

- What are the unique characteristics of each component in the integrated system?
- How do the characteristics complement one another?
- What physical interfaces exist among the components?
- What data and information interfaces exist among the components?
- What ideological differences exist among the components?
- What are the data flow requirements for the components?
- What internal and external factors are expected to influence the integrated system?

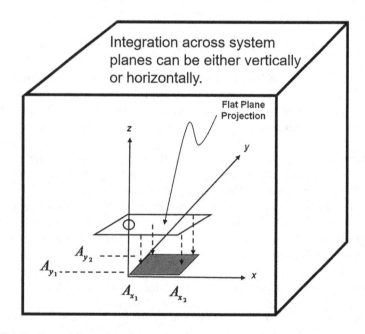

FIGURE 4.5 Systems integration geometric surfaces.

- What are the relative priorities assigned to each component of the integrated system?
- What are the strengths and weaknesses of the integrated system?
- What resources are needed to keep the integrated system operating satisfactorily?
- Which organizational unit has primary responsibility for the integrated system?

Other quantitative alignment and integration methodologies are available in the literature for consideration in this last stage of the DEJI System Model.

CONCLUSION

Outputs of SoS follow an integrative process that must be evaluated on a stage-by-stage approach. This requires research, education, and implementation strategies that consider several pertinent factors. This chapter suggests the DEJI System Model, which has been used successfully for product development applications, as a viable methodology for system design, system evaluation, system justification, and system integration. The integrative approach of the DEJI System Model will facilitate a better alignment of technology with future development and needs. The stages of the model require research for each new product with respect to design, evaluation, justification, and integration. Existing analytical tools and techniques as well as other systems engineering models can be used at each stage of the model. Thus, a hybrid systems modeling is possible. It is anticipated that this paper will spark the interest of researchers and practitioners to apply this tool in new system development initiatives. For example, in a contested political environment, an engineering systems approach can be applied to evaluate and justify a politically driven national proposal, with a specific template for integration into the expectations of the local population. This can help mitigate the common realizations of failed political promises. Such an application will expand the applicability of the DEJI System Model to unconventional platforms of national debates related to economic development.

CASE EXAMPLE: SHOWER CURTAIN POCKETS

This is a case example of assessing if a shower curtain needs pockets. It involves a defunct design of a shower curtain with pockets sewn on the bathtub side of the curtain.

Does a shower curtain need pockets?

Absolutely not.

The idea seems appealing and welcomed in concept, but the practicality of it is suspect.

When I first saw a plastic shower curtain at a residence, I told my wife this would not last. She thought it was a great idea.

Yes, the Design looks good. But, thinking further down the operational line, I knew that in usage, the pockets would not be functional.

First, users would have the tendency to stuff all manners of stuff into the pockets, thereby creating an impractical downward weight on the shower curtain rod. In no time the rod will sag and fall.

Second, blinded by shower water dripping down the face, users would fish around the curtain trying to find the right stuff stuffed into the wrong pocket, thereby upsetting the nicely draped shower curtain.

Third, over time, the pockets will start to fray at the sewn edges, thereby creating an unsightly look of the shower curtain.

Fourth, the process putting items into the pockets, retrieving them, and replacing them would, occasionally, create a messy spill of uncooperative liquids, such as shampoos, shower soap, hair conditioners, and so on.

Finally, opening the shower curtain to get out of the shower would be a weighty challenge because of all the stuff stuffed on it.

All these unpleasant and unwelcomed realities could have been discovered a priori if the DEJI Systems Model stages of Evaluation, Justification, and Integration had been in place.

An evaluation of how the pockets fit the décor would have revealed a disconnect.

A justification of why a pockets is needed and how often it would be used would have shown the impractical side of the shower pockets.

An integration review of how people actually use a shower curtain vis-à-vis the design would have highlighted operational misalignments.

REFERENCES

Badiru, A. B. (2012, Fall). Application of the DEJI model for aerospace product integration. *Journal of Aviation and Aerospace Perspectives (JAAP)*, 2(2), 20–34.

Badiru, A. B. (2014). Quality insights: The DEJI model for quality design, evaluation, justification, and integration. *International Journal of Quality Engineering and Technology*, 4(4), 369–378.

Badiru, A. B., & Racz, L. A. (2018, June). *Integrating systems thinking in interdisciplinary education programs: A systems integration approach.* Proceedings of the Annual Conference of the American Society for Engineering Education (ASEE), Salt Lake City, UT.

Valerdi, R. (2014). Systems engineering cost estimation with a parametric model. In A. B. Badiru (Ed.), *Handbook of industrial and systems engineering.* CRC Press.

5 The Postal System in the Supply Chain

INTRODUCTION TO THE POSTAL SERVICE IN THE SUPPLY CHAIN

The postal service system, commonly referred to as the "post office," in any nation is a critical part of the supply chain. This fact is often neglected, ignored, dismissed, or lambasted by the general public. The premise of this book is to both elevate and optimize the postal service system in the context of the general global flexible supply chain. The slogan of the United States Postal Service (USPS) slogan is "We deliver." Yes, indeed, the post office delivers. Moving and delivery of items constitute the core of any supply chain.

> Neither snow nor rain nor heat nor gloom of night stops postal couriers from the swift completion of their appointed rounds.

We must keep the supply chain moving, in a variety of ways. It is the position of this book that the Postal Service System plays a critical role in that process, in ways that might not have been previously recognized. The purpose of this book is to highlight the roles and benefits of the postal service in ensuring and advancing a flexible supply chain. Some of the ideas and methodologies presented can also ping the postal service system to explore new technology-enabled techniques of moving things, whether digitally, virtually, or physically. This could make the postal service system to be more appealing to their customers and observers, compared to the traditional public views and perception of the "Post Office.

> If the snail is moving, at least we know it is moving.
>
> —Adedeji B. Badiru

Flexibility in business, industry, and other enterprises connotes the ability to respond rapidly to developments in the supply chain system. The approach of this book is to use the industrial engineering framework to improve efficiency, effectiveness, and productivity in any supply chain, with the postal service recognized as core component.

LUSTER LOSS

There is no doubt that the traditional post office has lost its luster over the years. This is at least in the eyes of a small segment of the population. This is often expressed in frequent commentaries by customers, albeit not entirely accurate. As customers become increasingly sophisticated, they tend to be more discerning and, thus, more critical of service providers. It is really in the respective perspective of each

DOI: 10.1201/9781032620701-5

customer. Personally, as a frequent customer of the post office, I never share the same negative perspective that many ungracious customers express. I see every service provided by the post office as a necessary, desirable, and worthwhile competitive community-focused program. So, I am always grateful to the post office. In fact, I have always rented and maintained a local post office box in every community that my family has lived over the past five decades. To me, the post office is the anchor to the entire community. This is view that I have maintained and promoted ever since I was growing up in Lagos, Nigeria, where I cherished the little old ladies who operated postal services by selling postal stamps out of the outside windows of their rickety homes around the local neighborhoods. In fact, such postal outlets connected me with the world in the days that I was applying to overseas universities for further education. So, my appreciative perspective is different from those who never enjoyed the foundational roles of local post offices.

The loss of luster of the post office is not due to a lackluster performance in the stock market. Rather, it is due to the emergence of alternate modes of communication in community-to-community linkages. Thus, some customers are apt to be critical of what they see (or don't see) at the local post office. As a quasi-government entity, the post office cannot engage in flashy upgrades like the private industry can. Yet, the post office is subject to the same cost escalations that are rampant in the private sector. If we are not careful, we will peg our service expectations of the post office to the same level of what private industry can provide. From my perspective, as the author of this book, what we need to do more is to help the post office to have access to and use the same process-improvement tools and techniques that the private sector enjoys. That is the premise of this book, which attempts to bring the tools and techniques of industrial and systems engineering to the operations of the postal service system. This is an operational partnership that could create a better win-win relationship between the community and the post office.

HOW THE POST OFFICE CREATED AMERICA

Winifred Gallagher (Gallagher, 2016) gave an illuminating historical account of how the post office created America, from the days of the Pony Express to the modern time of digital communications. The history of the US Post Office is nothing less than the story of America. Out of all of America's founding institutions, the post office is the least appreciated or studied constructively. Yet, it has been and remains the most consistent major investment of the nation. It is hoped that giving the post office a set of noticeable pages in this book will spark more constructive interest in not only helping the post office to survive, but also to thrive. That race is on now. It is expected that leveraging engineering's problem-solving methodology can give the traditional post office more visibility, recognition, and support. Following the American Revolution, it was desired to have a centrally created institution to circulate news throughout the political and social systems of the nation. Thus emerged the need for the US Post Office. The team of George Washington, James Madison, and Benjamin Rush determined to devise a means to do just that. Their unique idea of the American post didn't just carry letters; it also subsidized the delivery of newspapers to the entire population. The postal system, as envisioned by America's early leaders, has endured as one of the few American institutions (public or private) in which citizens are treated as equals. The post office serves everyone, young and old, rich and poor, males and females, and elite or common. The US Post Office was created on July 26, 1775, by decree of the Second Continental Congress. Benjamin Franklin was the first postmaster general of the US postal system. He put in place the foundation for many aspects of today's mail system in the United States. The Pony Express is so closely linked with the US Post Office that the mounted courier, who was the post's insignia from 1837 until 1970, is mistakenly assumed to be a Pony Express rider, in the same way that the slogan "The mail must go through" is thought to be the post's motto.

Some key takeaways from Gallagher's book include the following:

- For many decades, the stagecoach ruled America's roads and carried its mail in the East and the West.
- Although short-lived, the Pony Express demonstrated America's can-do spirit and helped keep California in the Union.
- "Stagecoach" Mary Fields, born enslaved, was a popular figure who drove the mail by wagon in the Montana wilds around the turn of the 20th century.
- Postmaster General John Wanamaker was a brilliant merchant who revolutionized mail service, just as he had done in merchandizing.
- In 1902, Rural Free Delivery brought the mail to the homes of appreciative agrarian families in small towns around America.
- Postal Savings Banking encouraged family thrift by moving cash from under the mattress into the general national economy, which especially benefited underserved Americans of modest means.
- Postal savings customers could open an account with just $1, but could not invest more than $500, which was later raised to $2,500. They received 2% interest, which pressured private banks to raise their own rate to 3%, thus creating a competitive opportunity for residents.
- Some postal carriers operated horse-drawn wagons, which were later replaced by trucks.
- Postal carriers provided important public health services, such as weighing babies and keeping community records. This was particularly very important in rural communities with limited government-run health centers.
- During World War I, mail to and from troops abroad were transported by transatlantic ships, thus connecting families with their sons in the service.
- The US mail system provided crucial early support for the aviation industry through fast air mail service.
- Iconic airmail pilot nicknamed "Wild Bill" Hopson died at the age of 38 doing the dangerous job of delivering mail. What a sacrifice for mail delivery!
- Wars had long provided employment opportunities for women.
- The first female letter carrier in Chicago was Jeannette Lee in 1944.
- The proactive Postmaster General, Arthur Summerfield, appointed by President Dwight Eisenhower, initiated an attempt to modern the postal service through mechanization.

- In 1966, the rapid growth of postal services overwhelmed the postal workers in the antiquated Chicago post office, where operations ground to a halt.
- America's highly automated 21st-century post office handled 40% of the world's mail and provided the most productive mail service at the lowest cost.
- Postmaster General Montgomery Blair, appointed by President Abraham Lincoln, brilliantly used the exigencies of war to improve US mail service.
- Both the Union and Confederacy used stamps and "patriotic stationery" as political propaganda.
- The desire to afford privacy to recipients of bad news during the Civil War provided the impetus for mail delivery to urban homes.
- The *Chesapeake* was a mail steamboat.
- The mid-19th-century's postal services turned women into active correspondents.
- Star-Route carriers transported rural mail by the least expensive and most expeditious means.
- Mary Katherine Goddard was America's first woman postmaster.
- Disturbed by the railroads' high charges for transporting mail, Postmaster General Amos Kendall diverted to the old-fashioned post rider approach.

THE POSTAL WORKFORCE

The postal workforce is responsible for moving things around the world. Beyond the common public views of the post office, the postal workforce provides a critical service in the global supply chain. The postal workforce impacts flexibility and operational effectiveness in the supply chain, albeit behind the scenes most of the time. It is on the above notes that we must globally appreciate the marvelous and dedicated postal supply chain workforce, who are professionals in their own respective job functions. As articulated on the homepage of the Industrial and Systems Engineering Department at the University of Oklahoma:

> Industrial and systems engineers (ISE) design, enhance and manage complex, large-scale data-driven processes and systems to inform decision making. Involving both people and technology, ISEs help companies use resources effectively to solve complex problems. ISEs use computer-based tools to analyze information and present solutions, by combining engineering expertise with a business perspective. Companies seek ISEs for their expertise in understanding, evaluating and improving the performance of entire technical and business systems. Because of its versatility, an ISE degree opens doors for careers in business, health care, consulting, government and manufacturing.

Consulting and operational design and implementation are the readily common ways that industrial engineering could serve the needs of the postal service system. As it is in conventional industry, so it is in the postal service industry.

THE FIRST US POSTMASTER GENERAL

To recognize the postal workforce of the present, we must recognize where and how it all started. Benjamin Franklin was the first Postmaster General of the United

FIGURE 5.1   Original post office (Postamt) in the town square in Bonn, Germany, featuring the bust of Beethoven.

Photo credit: Adedeji B. Badiru, May 9, 2024.

States. He was a polymath, a leading writer, scientist, inventor, statesman, diplomat, printer, publisher, and political philosopher. Among the most influential intellectuals of his time, Franklin was one of the Founding Fathers of the United States, and a drafter and signer of the Declaration of Independence. In his diverse activities of his time, Benjamin Franklin must have been the first industrial engineer in his thinking, scheming, and acting. His ideas about the movement of goods, services, and information (e.g., mail) from one place to another represent the first flexible supply chain system, albeit within the limited scopes of the needs of that time. Since the time of Benjamin Franklin, the postal service (aka the post office) has contributed directly to the advancement of nations through the movement of information, often actuated as a postmarked letter. The post office has moved beyond carrying letters and packages. The service is now embedded in other services of moving general items. Vending of stamps, envelopes, boxes, greeting cards, and so on are more ways the post office has diversified into the global supply chain. The development of many nations was anchored to the availability and operation of post offices. As a boy growing up in several neighborhoods in Lagos, Nigeria, this author was always fascinated by the neighborhood-based postal outlets. That has left a positive impression and appreciation that have lasted until today. It was a big delight to see and visit (in person) one of the early post offices in Bonn, Germany, in May 2024, shown in Figure 5.1. The figure is a photo of the original post office in Bonn, Germany, featuring the bust of Beethoven (the acclaimed musician), who was born in Bonn.

COMMUNITY VIEW

My Community Love Affair with the Post Office: It is a delight to hear that the sense of reason has prevailed in the most recent political assault on our beloved post office system. The reversed decision of the postmaster general to defer proposed changes until after the 2020 election demonstrates where the hand of logic has touched an irrational development. The decision to make such a radical change in the post office should not have happened in the first place at this critical moment in the nation's battle of wits with COVID-19 and the shifty political landscape. Although many community voices were raised about the decision to make post office changes, I have not seen community-wide placards declaring "Don't mess with our post office." Through this *Dayton Daily News* medium, I hope more of the community will be sensitized to the plight of the post office, which is often lambasted for inefficiency and unprofitable operations. Beyond just processing mail from and to desired points within the community, many people don't realize the expansive value that the post office brings to the community. In any corporate setting, there are profit-generating entities just as there are non-profits. To constantly evaluate the post office on the basis of operational financial losses alone is to ignore the larger value of the post office as a community rallying point. The premise of establishing the post office in the first place was what was envisioned as the need for a community connection. Since its inception in 1775, the post office has dedicatedly provided that value. Whenever we consider the efficiency or profitability of the post office, we should always, concurrently, evaluate its value, with respect to the growing services to a rapidly growing and diverse population. This lack of value appreciation is where I often see a disconnect in the reactions to the lack of profitability of the post office. Many people don't realize that the post office is self-sustaining. Without dedicated national funding, the post office will always be fighting an uphill battle. Certainly, there is always room for improvement in any organization and the post has been doing its best. In my opinion, even though the losses may be mounting, I believe the value to the community is actually increasing. Unfortunately, value is not readily quantifiable. Thus, observers latch on to that which they can see and measure in terms of post office losses.

In more and more deep thinking about this issue, I have discovered that the reason I love going to the post office so much is the same reason that I am addicted to maintaining my consistent subscription to local newspapers wherever I have lived in the United States for over 45 years. That common reason is the community connectedness that the post office and the local newspaper provide. No matter how digital we take our operations, humans will still need human interaction and interfaces. Post office and local newspapers represent the consistent and reliable avenues for us to continue to be humans.

Telephone booths used to be a symbol of the community and a rallying point for all sorts of communications, both wholesome and illicit. But the advent of cell phones spelled doom for the phone booths, which are rarely seen these days. This should not be allowed to happen to the post office and the local newspaper. Community mailboxes should not be removed! Personally, I detest the recent promotional ads by the post office itself, encouraging customers to go digital and "never visit the post office again." Hogwash! That's like inviting us to stop being humans. The post office is the remaining, but thinning, thread of human connectivity in the community. It should not be allowed to be assaulted unabashedly by any political leanings.

FIGURE 5.2 Replica of a US postal truck given as my 71st birthday present.

Congress should fund the post office and help to increase its value to the community rather than letting it wither in the wilderness of modern sentiments of going digital.

So much so is my family aware of and enthused by my romanticism with the post office that for my 71st birthday in 2023, the coordinated present given to me by my family is a miniature replica of a US postal truck, shown in Figure 5.2.

POSTAL SYSTEM FOR NATIONAL DEVELOPMENT

In this section, we present the strategic pathway advocated by Alvar Bramble of Venezuela regarding using the postal system for national development and eradication of poverty. He presents specific cases of Mexico, Venezuela, and Russia. If this sounds far-fetched, it is because radical thoughts are needed for a radical national plight. In an October 1994 article, Bramble outlines the criticality of the postal system in national advancement efforts. His particular idea relates to Venezuela's IPOSTEL, Instituto Postal Telegráfico, which is a Venezuelan public regulatory organization of the national postal sector and provider of postal, logistic and printing services.

Contrary to popular belief, the primary function of mail services isn't communication between families, or between businesses and clients. Mind you, these are important features. In industrialized countries, business-generated mail stimulates commercial interchange and keeps delivery routes well-serviced by mailmen. The critical function of public postal services, however, is to help democracies work.

If this is true, then one can reasonably argue that only when IPOSTEL begins to provide high-quality postal services will the country's democratic institutions become stronger – and its public services in general more adequate. The main objective of a democratic system is to solve a society's problems by providing public services, dispensing justice, and protecting nations against internal and external dangers.

In modern nations, the media focuses on the major social issues and economic problems. Although less newsworthy issues don't make the headlines, it is precisely those problems that make up the bulk of a society's ills.

In Western nations, these everyday, seemingly insignificant problems, are often presented to legislators in a written form via the mail. A legislative cycle begins when problems are made public, either directly by citizens or through the mail. Afterwards, there is discussion in the legislatures so that problem-solving bills can be proposed and approved. The cycle ends with the discharging of laws by the executive branch.

The last event of the legislative cycle should solve the problem that was made public in the first place: This last event is what we call a public service. The dynamic nature of social life always creates new conflicts and new problems. So, cycles are continuously starting anew. This recurrence generates a legislative process-flow in a closed loop. Thus, a nation with a steady flow of information to its legislatures make for a country with a fully operative democracy.

But without adequate mail services, this vital loop is unable to close, and social problems go unresolved because they are never presented to legislators.

The minimal amount of factual information that congressmen do receive from constituents often results in irrelevant debates, incoherent laws, and the inadequacy of all law-mandated public services.

A good way to observe this gap in operation is to approach the Venezuelan Congress on any working day and notice constituents from all corners of the country trying to arrange a meeting with their congressmen. Mail campaigns would, undoubtedly, be a much more efficient way to take care of this sort of business. Instead, politicians' offices throughout the land are packed with citizens who must take time off work to present their views and problems to government. In reality, only a very small percentage of the public can make these types of visits. The problem is that the resulting lack of intercommunication makes the public assume that Congress is not aware of its need and is, therefore, a worthless institution.

This makes IPOSTEL's service possibly the most critical public service, since it is the only one capable of feeding the Venezuelan Congress with the type of information citizens have regarding social problems. Congress needs that information to make the Venezuelan legislative system achieve its objectives: That is, to provide quality public services to the Venezuelan society. If the Congress is to gain the respect and trust of the people that it deserves, IPOSTEL must begin to facilitate inexpensive and efficient correspondence between all constituents and their congressmen. Modern postal services can also make a big difference in daily government activities such as tax collection and traffic control. An efficient IPOSTEL, for example, would permit police to clamp down on reckless driving, since through the auspices of a reliable postal system the courts could track down traffic violators at home using the mail. This would instill respect in general for traffic laws, and indeed for all laws in Venezuela. But what about the case of laws that are unjust or incoherent? An adequate postal service empowers citizens to express their dissatisfaction by demanding legislative changes and even accountability from their public officials. Accountability would result in laws reflecting social needs. But accountability can only be obtained if citizens can follow up on the work of elected officials through an up-to-date postal system.

By closing the existing communications gap with quality mail services, Venezuelan leader would be able to understand the full extent of the country's problems and come up with solutions to its many social ills. In developed countries. postal services are among the largest and most visible of government employers. This is not the case in Venezuela. Yet the size and efficiency of a nation's postal service can make the difference between a rudimentary legislative system and one that is able to unleash the full potential of democracy. For these reasons. Venezuela's leadership should commit itself to making IPOSTEL a strategic component of the nation's democratic system.

Recap:

Obviously, the nation of Venezuela did not follow up on Bramble's recommendation above, as the recent profile of the nation has degenerated into an economic chaos and an unstable political system.

Bramble also presents his postal ideas about Russia. The case of Russia is not unlike that of many developing countries, where there is a dichotomy of national world-power profile side by side with miserable plights of the poor. Yes, Russia is a "world power," so to speak, militarily. Yet, a large block of the Russian population is subjected to abject poverty. Bramble captures the essence of this dilemma in the contest of leveraging industrial engineering and the postal system to address and redress national development challenges.

In 2024, Bramble writes on "The Eradication of Poverty in Russia: An AI-Driven Participatory Method for Meeting the SDGs." It should be recalled that the postal service is a participatory system. Using the UN's Article 21–1 of the Universal Declaration of Human Rights as the framework, Bramble echoed the following quote:

Everyone has the right to take part in the government of his country, directly or through freely chosen representatives.

(United Nations General Assembly resolution 217 A (III), December 10, 1948)

Where citizens have the freedom of choice, the participatory context of the postal service provides a foundation for national development.

The industrial engineering methodology employed by the Global SoS Network (in this document, the Foundation) has identified the lack of adequate postal services as a logistical constraint that impedes efforts to eradicate systemic poverty both in Russia and around the Global South. Indeed, the Foundation has found that most major challenges facing both the Russian Federation and the Global South are systems problems, not people problems, and that the same aforementioned logistical constraint, in the form of inadequate postal services, is the cause of such systems problems.

When postal services are adequate, utilities and infrastructure managers can operate effective billing services. When that occurs, the quality of public services is also adequate because there ample funding for delivering quality services. Under those conditions, business managers become motivated to invest in technology that will improve productivity. The resultant gains in productivity yield increases in income, which ultimately reduces poverty.

Fortunately, in March of 2013 Vladimir Putin took a required first step to resolve the Russian situation by becoming the first 21st-century global leader to decree postal services as a national strategic activity. At the moment, Putin did initiate a postal service improvement program in Russia. But the initiative has largely stalled. Consequently, Russian Post (Russian: Почта России, Pochta Rossii), does not yet operate a postal service that will help Russia reap the full socioeconomic benefits that Putin's initiative should have provided by now.

Russia additionally endures the effects of a technological constraint that causes, for example, localized poverty in addition to causing most of the other challenges – including environmental degradation. This second constraint consists of the general lack of knowledge about the importance of communicating citizen concerns and opinions to representatives in government.

Nonetheless, Russian federal law calls for relationships between citizens and their representatives in the Russian Duma, and as with all matters of the state, the relationship should be established formally, in writing. This has not been occurring precisely because of the lack of postal services adequate to transport documents between Russian citizens and their representatives in the Duma.

Technical progress allows for the use of AI-enabled constituent dashboards. These would act as facilitators in the process where Russian citizens uphold Russian federal law by establishing relationships with their legislators in the Duma. Consequently, AI technology offers improved possibilities for Russian constituents to influence Russian legislators towards the delivery of wellbeing and improved productivity for all Russians.

SUBJECT: This document describes how to use AI so that Russia can eradicate poverty while meeting the SDGs, and serve as catalyst for the same process around the Global South.

OBJECTIVES: The objectives of this document are:

• Demonstrate the sociotechnical processes required for making steady gains in productivity in order to eradicate poverty in Russia;

- Identify the fundamental causes for poverty in Russia and the Global South;
- Recommend practical solutions.

Background

1. As leaders of The Russian Federation seek methods to resolve the poverty problem in Russia, it is important to note that a lot of the money being employed around the world towards the elimination of poverty is wasted because there has been no agreement on what is the best thing to do, or "best practice."

2. The United Nations has established 17 Sustainable Development Goals, with the critical number one goal of eliminating poverty by the year 2030. The usual methodology to reduce poverty has been to employ the reductionist approach, which originated in the London School of Economics after WWII. The approach is based upon the belief that interventions into the various facets of poverty will eventually overcome and gradually eliminate pockets of the problem. And as the manifestations of poverty become eliminated one by one, eventually the reductionist approach will eliminate the overall problem. However, the methodology has evidently has not been successful because poverty is firmly entrenched in too many countries of the world.

3. In some quarters, it is generally known that improving the quality of public services plays a role in the reduction of poverty. In other quarters it is also known that the delivery of public services is a very expensive undertaking, especially in rural areas because of the distances involved. In yet other domains, such as the work environments for utilities and infrastructure managers of developed countries, it is common knowledge that billing systems operate best through the formality, convenience and economies of postal services, which enable the mobilization of local resources.

4. From the point of view of Operations Research/Management Science (OR/MS), the postal services of a nation can be considered to operate in two major domains. Those components are *internal plant* and *external plant*. The *internal plant* component of a postal administration comprises all areas that usually house office space and equipment, within the perimeters of all post office buildings and sorting centers.

   Alternatively, the *external plant* is the external postal infrastructure, out in the streets. It includes public mailboxes where citizens place their mail for delivery, and the addresses and artifacts required to make postal deliveries. These include street nomenclature, address number, and private or family post boxes where the final deliveries are made by postal delivery personnel.

5. According to information retrieved from Internet news sources, there have been concerns that Russian Post offers an unreliable service. Nonetheless, in 2013 the president of The Russian Federation, Vladimir Putin, decreed Russian Post as a strategic enterprise, and initiated a postal service development program. Afterwards, Russian Post contracted an Italian company to

help make improvements to the Russian postal service. However, the scope of the project excluded the *external plant* of the Russian postal service.

6. The Universal Postal Union (UPU), a UN body located in Berne, Switzerland, unifies the postal administrations of all nations. The UPU helps countries develop their postal services, and it has given the name Postal Delivery Point (PDP) to the final destination where postal workers are to make final delivery of correspondence. Each PDP represents a home which should be equipped with a family mailbox or postbox. It is the exact location where postal workers will deposit the correspondence they have transported to the delivery point.

7. As Russia seeks to improve its international standing, it has provided Official Development Assistance (ODA) to several nations, however, given the experiences and historically uneven results of ODA, it is safe to assume that many of those resources, including resources expended by the UN SDG Action Campaign, in Bonn, Germany, have not been put to effective use.

8. As Russian industry seeks new markets, stagnant productivity and widespread poverty are a hallmark of the early 21st century in Russia and the Global South. Since most countries of the world are poor, the solution of the poverty problem will probably create giant new markets.

9. The sociotechnical approach to public affairs relies on the concept that nation-states are made up of social and technical components. The social component of nation-states is society at large, and the technical component is the machinery of government. The sociotechnical approach holds that wellbeing and productivity are a result of harmonic interaction between social and technical components.

   The sociotechnical approach expresses that if a sociotechnical system does not deliver satisfactory results, it is usually because of lack of harmony between the social and technical components, and steps are usually taken to obtain joint optimization. Usually, changes are made to the technical aspects in order to optimize the outputs of the overall system.

10. Furthermore, software technical progress has brought about artificial intelligence technology, which promises to help revolutionize the way information and communication is managed by individuals and organizations.

Findings

1. According to operations research (OR) carried out by the Globalsosnet foundation, there was a need to employ the Theory of Constraints to identify the causes of poverty in Russia and the Global South. The results indicate two major poverty-creating constraints in Russia and around the globe:
   - A logistical constraint
   - A technological constraint.

2. The largest poverty-causing constraint is logistical, and it consists of the lack of capacity for mobilizing local resources. In Russia it originates at the *external plant* of Russian Post, and its most visible characteristic is the lack of sufficient functional Postal Delivery Points (PDPs). In other words, in Russia there are many homes not equipped with functional mailboxes.

Mailboxes are probably the most important component of the national infrastructure because they are key points for the financing of the state. Mailboxes are the access point through which tax authorities, utility and infrastructure administrators can access financing streams originating from taxpayers or from clients. When a home is not equipped with a functional mailbox or postbox, it means that it is a non-functional PDP. It also means that the tax authority or service provider cannot count on the income required to deliver expected results.

On the other hand, PDPs are the central point of attention in the working environment for postal delivery personnel. Postal delivery workers complete a task only when they deliver mail to a PDP at a home. However, when family post boxes are in a state of disrepair or missing, postal delivery personnel cannot make efficient final deliveries. Unstable or not fixed in place, rusty or unkempt PDPs lower postal worker morale and motivation, blocking the effective delivery of mail and reducing the reliability and efficiency of the whole Russian Post organization.

The consequences are dire, because the situation impedes the operation of effective billing systems by utilities and infrastructure managers. Without effective billing systems, there are no income streams capable of financing quality public services, or capable of financing the development and upkeep of the public infrastructure – especially in rural areas because long distances involve higher operating costs.

This situation causes systemic poverty because low quality public services place psychological constraints on business managers. When that is the case, business managers do not become motivated to invest in technology that would improve productivity. This makes it difficult if not impossible to raise wages without raising the sale prices of the products or services offered by businesses, so the standard of living stagnates at a low level and the country stays poor, which is the usual case in Russia and the Global South.

When good public services are the norm, business managers do invest in technology to improve productivity and can then increase wages without having to increase the sale prices of their products or services. In addition, with adequate postal services, productivity and the standard of living usually make small yearly gains, and the country undergoes stable economic growth along with continuous reduction of poverty.

In that light, during March of 2013, President Vladimir Putin decreed Russian Post as a strategic enterprise, but the postal development program Putin ordered focused mainly towards internal plant. This means that the new sorting and transport equipment can process a volume of postal correspondence that the external plant infrastructure probably cannot handle in its present state.

3. The second poverty-causing constraint in Russia is technological. It creates localized urban and rural poverty, and it exists because there appears to be a general lack of awareness among Russian citizens regarding the importance of communicating citizen views and opinions to representatives in government.

Even though Russian federal law specifies procedures for such communication, the technological constraint is powerful because Russian legislative decision-making processes appear not to have a continuous input from citizens. In consequence, when public officials make decisions, they are not necessarily the ones required to resolve the concern or problem at hand, ranging from the quality of postal services to other problems at the local, regional and national level.

This explains why there is a host of conflicts, pervasive alcoholism, low productivity and relatively high poverty levels in Russia. The aforementioned lack of adequate postal services worsens the technological constraint because if the mail service is unreliable, people might find it complicated and even far-fetched to contact their representatives in government in regards to a specific problem, and might even think that it is improbable to get a reply at home from their representatives.

4. The use of AI-enabled dashboards creates a paradigm shift because it enables citizens, in their roles as constituents, to communicate grievances and opinions to their representatives conveniently and efficiently. The role of the dashboards would be to institute and expand relationships between constituents and their deputies in the State Duma. The dashboards would help Russian constituents grasp and better comprehend relevant issues. AI-enabled constituent dashboards can also help improve the writing abilities of Russian constituents by helping them redact concise but comprehensive letters to their legislators.

5. The first objective of a constituent is to obtain a reply. Once Russian constituents start to use constituent dashboards, they can – when necessary – exploit the existing logistical constraint by using the Russian postal service infrastructure to obtain their replies from contacted deputies at the state Duma. They can pick it up at the local post office through the Poste Restante (General Delivery or GP) service. The use of Poste Restante service completes the communication cycle for constituents. And each time a Russian constituent receives a reply from a deputy at the Duma, the constituent knows that the deputy has read and understood the original letter sent by the constituent.

6. The Russian Federation has already taken very important steps to resolve the technological constraint. Russian federal law describes some procedures for creating interaction between voters and their deputies in the State Duma, so in this context it will be straightforward to eliminate the technological constraint because it is just a matter of upholding Russian federal law. This makes it easier to visualize how high volumes of letters between constituent and legislators will create greater accountability in Russia, and in other countries once implemented there. Although Russian law refers to citizens residing in a Russian constituency as voters, in this document, the Foundation classifies and refers to them as constituents.

The applicable legal procedure appears in Article 8 of the Federal Law No. 3-FZ dated May 8, 1994, "On the Status of a Member of the Federation Council and a Deputy of the State Duma of the Federal Assembly of the Russian Federation," which deals with the "Interaction between a deputy of the State Duma and voters," and establishes, for example in its Article 8–2, that "A deputy of the State Duma must consider appeals from voters and should meet periodically with voters (constituents)."

In addition, Article 21–1 of the UN Universal Declaration of Human Rights is also applicable because the article states that everyone has the right to take part in the government of his country, directly or through freely chosen representatives.

7. The postal improvement program initiated by Vladimir Putin, and contracted with an Italian company, appears to have been focused towards the *internal* postal infrastructure. The result has been that today, even though Russian Post may have the latest postal-processing and transport technology, the lack of maintenance and upkeep of the *external plant* maintains the overall postal service at an inadequate level.

   This makes it difficult for Russia Post-delivery personnel to deliver and deposit the mail at each PDP in an efficient manner. Because they are components of the relevant environment for postal employees, unkempt or missing post boxes cause stress on postal delivery personnel and probably result in random delivery events: The postal worker may or may not deliver the correspondence to each assigned PDP.

8. In spite of the existence of Internet-based billing systems, the event makes it problematic for utilities and infrastructure developers and managers to operate effective billing systems and then to invest and maintain the infrastructure. As explained before, this situation limits investments and limits productivity, which in turn increases poverty.

9. Furthermore, because the use of AI-enabled dashboards increases the use of postal services, AI is a factor that enhances domestic unity and national cohesiveness.

THE GLOBAL POSTAL DIVIDE

The lack of resources is the most visible landmark of poverty in the world, and in order to eliminate poverty, the mobilization of local resources has been identified as one of the objectives to be met by the UN Agenda for Sustainable Development (SDG 17.1). But the mobilization of resources requires physical means to transport those resources, and postal services that are adequate for the task are the most convenient and economical services that can provide such transport.

The Foundation has found that the fundamental reason Global North countries are rich is because the volumes of mail that these Postal Services transport make them adequate for mobilizing local resources. On the other hand, postal services operating in Russia and the Global South do not transport letters in sufficient numbers that would be adequate for the mobilization of local resources. Consequently, the low quality of postal services in Russia and the Global South impedes the creation of wealth, which places these nations under a global postal divide.

1. According to postal statistics reported by Russia Post to the UPU, Russia Post delivered about 16 letters per inhabitant in 2017 (items 2.3 and 1.2 in the database for Global or regional estimates). However, postal services

in rich countries delivered at least 40 letters per year per inhabitant. As an illustration, the postal delivery rate for Canada is about 142 letters per year per inhabitant. Reference: www.upu.int/en/Universal-Postal-Union/Activities/Research-Publications/Postal-Statistics

- In contrast, poor countries usually have a delivery rate of less than 5 letters per year per inhabitant. For example, Switzerland has a postal delivery rate of 256 letters-yr./inhab., while for Zimbabwe it is 0.38 letters-yr./inhab. These results indicate that it is a lot easier for Swiss utilities and infrastructure managers to bill their customers regularly. The results also offer an answer as to why public services are good in Switzerland, and why there is a good investment climate along with high productivity and high salaries. But the statistics for Zimbabwe indicate that, in that country, billing for utilities and public services is very likely a difficult endeavor notwithstanding the possibilities of Internet billing. It also gives a good indication why the country has abysmal public services and consequently, systemic poverty.

    While statistical regression analysis of the data will probably establish correlation between postal deliveries rate and systemic poverty levels, there are clear indications that merit the establishment of causation. When the number of postal deliveries is relatively low, it is probable that municipal and utility billing systems are not able to bill most or all their services, and are unable to help maintain the income stream required for the improvement of public services and the infrastructure.

- For instance, the Globalsosnet Foundation estimates that in Russia there are around 52 million households that should each receive monthly billing for water, electricity, gas, local taxes, building maintenance, and perhaps telephone use. The Foundation also estimates there are roughly 10 million business and other establishments that should also be receiving around 5 monthly bills. This means that around 60 to 70 million recipients, at a rate of around 5 bills per recipient, should be receiving around 350 million monthly bills or 4.2 billion bills per year. But according to Item 2.3 and 9.2 of the database, the global delivery rate for Russia is around 1.9 billion letters per year per inhabitant. Because postal services usually deliver other mail besides billing, this number does not seem to be adequate for delivering even a quarter of all the utility and other public service bills – and tax bills – that should be delivered every month in Russia.

    A formal statistical analysis of the situation can reveal how many bills are being delivered by Russia Post every month, and very importantly, how many bills are not being delivered by the postal service because of faulty PDPs. Each undelivered bill lowers the existence of available resources and illustrates the urgent need for improvements in this area.

2. The same photographs also help make a very important revelation. For economic success, productivity is paramount, and if Russia Post delivery personnel are to be really productive, they must deliver the mail as efficiently as practically possible.

In order to improve postal-service efficiency in Russia, family names should be placed on the *outside* of every family postbox, in public view, so as to identify the post box publicly, and help the post-delivery personnel not waste time identifying recipients, and thus make quicker deliveries. Among the benefits of this practice would are easier training of postal delivery workers, fewer delivery errors, and less returned mail. The identification of family post boxes with family names results in more effective billing processes, and more resources to eliminate the logistical constraint along with systemic poverty in Russia.

3. The 21st-century practice of billing through the Internet is attractive and necessary because of the allure of instantaneous communication, but definitely not sufficient for satisfying the billing system requirements of public services. It can then be deduced that if Internet billing were sufficient, low productivity and high levels of poverty would not be such large problems. Reference: http://globalsosnet.cfsites.org

   In addition, for billing of utilities and taxes, billing through Internet does not present the legal formality nor the proper deference of public servants to citizens. If the service provider or tax authority bills by sending e-mail or text messages, a citizen could surmise that if the bill is not so important as be sent formally, signed and sealed. Why should it be important to give it priority? Yet again, this situation further diminishes the income stream that authorities need to invest in the infrastructure required for development.

4. Although part of the UPU's job is to help countries develop their postal infrastructure, it seems to be carrying out the task on a limited basis. As of the year 2023, countries that the UPU has already helped develop such infrastructure for example, Costa Rica and South Africa, do not yet operate what could be considered to be adequate postal services because they do not deliver the volumes of mail for making them adequate postal services.

5. The Foundation has also found that there is a causal chain between the lack of adequate postal services and a large number of problems such as conflicts, crime, and poverty existing in Russia and the Global South. All these are precisely the situations that the SDGs were designed to resolve, and most stakeholders agree that broad public participation is the best way to resolve such a large number of problems and conflicts. And because the elimination of the constraints relies on citizen inclusion and engagement with their representatives in government, AI helps visualize a sociotechnical procedure to create the conditions from which the achievement of the SDGs would be emergent properties. It is an effective and very democratic manner to resolve all the conflictive situations, and to eradicate poverty.

THE SOCIOTECHNICAL METHOD

In order to eradicate widespread poverty in Russia the following method should be used, and then scaled up to the Global South:

- As a means to uphold Russian federal law, develop and roll out AI-enabled constituent dashboards for computers and smartphones.

- Assure that dashboards have been designed with the operational objective of sending of messages or letters to deputies in the State Duma.
- Install constituent dashboards in post-office Internet kiosks so that constituents can grasp issues and redact letters to be sent to deputies in the State Duma.
- For constituents who presently do not receive mail at home, distribute instructions for encouraging them to use the Poste Restante (GS) service as their return address. This step helps assure that constituents who contacted their deputies will get formal replies.
- Initiate "Write to your Deputy at the Duma" institutional campaign.
- Once rollout of dashboard has initiated, consult with deputies at the Duma regarding the installation of mail management systems, for the management of increasing volumes of messages and letters from constituents.

1. Uphold Russian federal law in order to break the technological constraint.

Constituent dashboards should be developed for use with Russian-language software in Russian computers and smartphones so that Russian constituents can focus on establishing correspondence with their deputies in the Duma. The rollout of constituent dashboard applications would allow voters to interact and establish relationships with their deputies, as per Russian federal law that deals with the "Interaction between a deputy of the State Duma and voters." Consequently, and following Russian law, Russian constituents would use AI-enabled constituent dashboards, at home or in post offices, to help them prepare concise but comprehensive letters to their deputies in the Duma.

At the post offices, geographic constituency locators with the addresses of local, regional, and national legislators representing the area serviced by the post office should be prominent features. Constituent-friendly constituency maps and artificial intelligence information should be posted on screen savers. This information should be a task facilitator for constituents – or participatory planning teams of constituents – who might want to have a better grasp of relevant issues, or who need more information about their legislators, or want to interact with other constituents of the same or other constituencies.

A key component of the AI-enabled-dashboard rollout could be the promoting of best practice through "Write to your Deputy at the Duma" institutional campaigns. Consequently, constituent would not be passive spectators waiting for their deputies to meet them.

The use of AI offers convenience and meaningfulness because AI-enabled constituent dashboards would help citizens express their opinions and suggestions to their deputies in the State Duma. In addition, when AI mail management systems help deputies manage high volumes of communications and then help in preparing and sending replies. Prompt replies offer meaningfulness and satisfaction to constituents. The result will be systematic communication between Russian constituents and their legislators in the State Duma.

2. When constituents list the local post office Poste Restante service as their return address until postal service at their homes is resolved, they will obtain their replies from legislators by visiting the post office and inquire about replies from Duma deputies. The fact that Russian postal services are inadequate does not exempt the use of postal facilities like post offices and their services.
3. When constituents express dissatisfaction because the quality of postal services impedes communication with legislators, the latter will sponsor and encourage the improvement of the quality of postal services because the objective of legislators is to obtain votes for reelection. As mentioned, such improvement in the quality of postal services is key for improved investments, for gains in productivity, and for reduction of poverty.
4. The Russian government and the members of United Nations should look into the reasons behind the lack of capacity that the UPU has had in order to help developing countries improve the quality of their postal services as means to reduce poverty. With the participation of an effective Universal Postal Union, Russia should assist neighboring and other developing countries, first in eliminating their logistical constraint. Subsequently, and with the added participation of the Inter Parliamentary Union, assistance would be offered for eliminating their technological constraint. These activities will set the stage for those and other countries to meet the SDGs.

Also, the UN SDG Action Campaign, in Bonn, Germany, will need to receive assistance and collaboration with experiences on how all countries, both underdeveloped and developed, have been developing SDG activities, and how they can help nations exceed the SDGs at a fraction of the previously projected cost.

NARRATIVE OF NATIONS

By leading the efforts of employing the scientific method for eradicating global poverty, The Russian Federation can become the first country to meet all the SDGs, which in effect will mean the eradication of poverty in that nation. And by helping other nations implement measures for breaking both their logistical and technological constraints to sustainability, Russia can gain business opportunities that will lead to growth in manufacturing capacity and advantageous entry to giant new markets. This goes without mentioning that postal services themselves have a large job-creation capacity.

For the year 2013, the European Commission informed that postal services employed 1.2 million people in Europe. The effect of eliminating the logistical constraint in Russia and other countries will be to greatly increase that figure because adequate postal services stimulate business transactions. There will be ample and increased demand for design and manufacture of postal sorting and transport equipment, and demand for billions of public and home post boxes for worldwide distribution.

Because demand for paper will also go up sharply, sustainable forest management, wood-processing and paper-making machinery will be necessary. In addition, there will be a growing need for postal-processing building construction services, trucks and other vehicles, and the development and installment of new information and communication technologies for the management and operation of world-class postal services in Russia and around the globe.

THE URGENCY FOR STRATEGIC CHANGE

Putin did act upon postal services as strategic, but in order to complete the job the international community still needs to help upgrade the *external plant*, that is, the external postal infrastructure. These photographs illustrate the Reengineering of the Russian Postal Service as consisting of upgrading and satisfying maintenance requirements for eliminating faulty Postal Delivery Points in Russia. The results would include a sharp increase in postal throughput, along with emergent unity of purpose. The job will probably call for further postal-service reengineering work in neighboring countries and the Global South.

It is cost-effective and clear-cut action that will result in (1) eliminating the logistical constraint and "Strengthen domestic resource mobilization . . . to improve domestic capacity for tax and other revenue collection." UN Sustainable Development Goals Target 17.1; (2) the eradication of systemic poverty in Russia and whichever Global South country the reengineering task is carried out; and (3) growth in manufacturing as a result of increased demand both nationally and internationally from the postal service and other industries.

For example, if the address number falls off the façade of a building in Siberia, there will not be effective postal deliveries. Situations like this affect the performance of Russian Post, lowers investment that would increase productivity, and increases poverty. Faulty apartment building mailboxes in Moscow, Russia, make it difficult for the Russian Post to deliver monthly billing and degrade citizens' connectivity to the national pie. Home mailboxes are probably the most important components of a national financial system. They facilitate and allow families to share the cost of running a successful state. They also empower the state to fruitfully pursue targets such as the SDGs.

In spite of being a global power militarily, the many pockets of poverty around the Nation create an impediment to the level of national develop fitting a global power.

POST OFFICE TESTIMONIAL

Regarding the postal engagement with the community discussed earlier, one important role of the postal workforce is engaging with the possible future generation of the workforce through community communication with youths. Figure 5.3 shows postal worker, Kathy Cretella, engaging with young people about the community service of the postal system. In a community-connection testimonial, Kathy Cretella, outlines what the post office means to the community:

> I'd always heard good things about the post office.
> (Kathy Cretella, retired letter carrier,
> https://mountvernonnews.com/stories/637514127-local-letter-
> carrier-on-job-i-d-always-heard-good-things-about-the-post-office)

FIGURE 5.3   Postal worker on the job communicating with young children.

In a 2022 newspaper account, Gregory Burnett narrated a conversation with Kathy Cretella about the importance role of the post office in any community. The conversation goes as follows:

Postal carriers are still at the heart of American life, despite the times we have received our neighbor's mail or waited until sundown to see our boxes filled.

Imagine not having the convenience of sitting at home and sipping a beverage while waiting for an important document or heartwarming greeting card to be delivered personally.

That's what makes Kathy Cretella's job so important, and she realizes that. Cretella is a United States Postal Service letter carrier in Mount Vernon. She has been on the job for over 15 years. She started her career as a window clerk in a Montana town near Yellowstone National Park.

"At the time, I was doing a lot of odd jobs in Montana," Cretella said. "I was doing a lot of piecemeal jobs because I was a stay-at-home mom and my daughter was in school. And that's when I decided to apply for the clerk job and the post office. I'd always heard good things about the post office. So, I wanted to give it a try."

After moving east to Mount Vernon, Cretella was able to transfer to the town's postal station as a letter carrier. She is one of two women who deliver mail in Mount Vernon. Cretella delivers in the neighborhoods and her partner, Amy McFadden, covers the downtown area.

"I'm on the eastside," Cretella said. "I do have some businesses. But, McFadden's route handles a lot of companies in downtown Mount Vernon. I walk about 12 miles a day through Mount Vernon neighborhoods. We are fortunate that the USPS gives us a clothing allowance yearly because I go through a sturdy pair of New Balance shoes twice a year."

Cretella is married and has a 28-year-old daughter. They live in Mohican State Park.

Technology has taken a big bite out of the postal service, like other professions. Many people prefer to send a text message with a heart emoji attached than taking the time to write a note inside a card.

"I've seen a big decrease in letters and cards," Cretella said. "So, when I see letters and cards in my routes, I am thrilled that people are still using the post office to further the written word. The parcel has increased by tenfold since I first started."

Cretella's shift typically starts at 7:30 a.m. She is out the door on her route by 8:15 a.m., returning to the station at 4 p.m. to check out.

Some people think the biggest challenges when delivering mail are snow, rain, wind, and cold. No; according to Cretella, it's dogs.

"Dogs are always an issue for letter carriers because they bark when we come, and when they bark, we go away," she said. "Here's an interesting tidbit: If we get bitten by a dog, which I have, we are told it's our fault. This job required you to walk and chew gum at the same time. You must be aware of your surroundings. The best part of my job is getting paid to exercise, and I love meeting the people on my route."

Cretella hopes more young people will want a career in the postal service.

"It's a feeling of tremendous satisfaction when you start with a huge pile of mail and ending [sic] the day knowing you did something productive," she said.

PROCESS IMPROVEMENT OF THE ROYAL MAIL

Bendell et al. (1993) presenting a very constructive case study of a process improvement strategy for the Royal Mail in the United Kingdom. Applying the measuring and monitoring techniques of quality management, the authors describe a Total Quality Management approach to improving the handling and delivery of Royal Mail. Royal Mail is one of the three main businesses of the UK Post Office Group. With an over 350-year history, the Royal Mail needed a modern transformation in the 1990s. The first step in the change process was the agreement of the principles behind the Customer-First direction for the business. Central to the process was the involvement of employees from all areas of the Royal Mail. In 1988, about 120 managers were brought together in small groups of six to define the current state of the business and the desired end state for the future. Several of the managers were initially reluctant to open up, believing that it was best to keep their heads down. Eventually, through positive and constructive presentations of the improvement plan, everyone got excited about the positive prospects. A list of shortcomings of the current situation was produced. Following further discussions, a list of requirements for the future was established. These include the following:

- A well-structured product range based on market research of customer needs;
- Effective measurement of customer satisfaction and a framework for action;
- Understanding by employees and customers of all products and services;
- Customer confidence generated by the presentation of a professional image in everything from uniforms and vans to enquiry offices;
- Reliability in meeting the performance specifications of each product, based on the needs of customers, without failure.

Based on the outcome of these discussions, and reflecting the requirements of the four stakeholders (customers, employees, shareholders, and community), the Royal Mail Business Mission was developed along with a statement of the values, which the business holds, concerning the care of customers and employees. Based on the multifaceted collaborative and inclusive efforts, what emanated was a comprehensive way of working throughout the organization, which allows all employees, as individuals and as teams to add value and satisfy the needs of their customers. Further, a business-wide

FIGURE 5.4 Royal Mail truck at Buckingham Palace, London, England.

customer driven strategy of change was developed, which moves everyone progressively to an environment of a steady and continuous improvement of everything that is done by the Royal Mail team. This case study is a great template for improving the postal system in any national development grid. It fits the theme of using industrial and systems engineering tools and techniques to improve the postal service as a part of the overall global supply chain. As an author, having earlier read this Royal Mail case study, I was enthused to have an opportunity to visit Buckingham Palace in 2023 and observed the operation of the Royal Mail delivery van (Figure 5.4).

REFERENCES

Bendell, T., Kelly, J., Merry, T., & Sims, F. (1993). *Quality: Measuring and monitoring*. Century Business Publishers.
Gallagher, W. (2016). *How the post office created America*. Penguin Books.

6 Modeling for Supply Chain Optimization

POSTAL OPTIMIZATION

Montazer and Shahbaz (2023) present a good example of the application of industrial engineering optimization modeling of the post office. Industrial Engineering and Operations Research (IEOR) is a discipline mostly formed in the 1900s and are young when compared to the longevity of the United States Postal Service (USPS), which was first established at its initial form in 1639. Interestingly, it has been observed that the IEOR tools and techniques of the current time have played major roles in the evolution of the USPS as we know it today. The purpose of this study was to track the history of USPS as it evolved, improved, expanded, and served the people of the United States, from 1600 to 1950, through the lens of the IEOR concepts, tools, and techniques. It has been observed that the USPS has gone through a vast number of changes, expansions, upgrades, adoption of various technologies to improve the quality of service rendered to the people of the United States, in terms of time, speed, reach, cost, and safety and security. In a closer examination, we also observed that some basic IEOR, though not existed in name, have been the tools and techniques instrumental in the evolution of the USPS through 1950. The USPS evolution beyond 1950 is being researched as a follow-up study and will be presented in future venues.

Schedule optimization is often the major focus in project management. While heuristic scheduling is very simple to implement, it does have some limitations. The limitations of heuristic scheduling include subjectivity, arbitrariness, and simplistic assumptions. In addition, heuristic scheduling does not handle uncertainty very well. On the other hand, mathematical scheduling is difficult to apply to practical problems. However, the increasing access to low-cost high-speed computers has facilitated increased use of mathematical scheduling approaches that yield optimal project schedules. The advantages of mathematical scheduling include the following facts:

It provides optimal solutions.
It can be formulated to include realistic factors influencing a project.
Its formulation can be validated.
It has proven solution methodologies.

With the increasing availability of personal computers and software tools, there is very little need to solve optimization problems by hand nowadays. Computerized algorithms are now available to solve almost any kind of optimization problem. What is more important for the project analyst is to be aware of the optimization models available, the solution techniques available, and how to develop models for specific project optimization problems. It is crucial to know which model is appropriate for which problem and to know how to implement optimized solutions in practical

DOI: 10.1201/9781032620701-6

settings. The presentation in this chapter concentrates on the processes for developing models for project optimization as presented by Badiru and Pulat (1995).

SUPPLY CHAIN MATHEMATICAL FORMULATION

Several mathematical models can be developed for project scheduling problems, depending on the specific objective of interest and the prevailing constraints. One general formulation is

$$\text{Minimize}: \{\max_{\forall i}\{s_i + d_i\}\}$$

$$\text{Subject to}: s_i \geq s_j + d_j \quad \text{for all } i; j \in P_i$$

$$R_k \geq \sum_{i \in A_t} r_{ik} \quad \text{for all } t; \text{ for all } k$$

$$s_i \geq 0 \quad \text{for all } i$$

$$r_{ik} \geq 0 \quad \text{for all } i; \text{ for all } k$$

where
  S_i is the start time of activity i
  d_i is the duration of activity i
  P_i is the set of activities which must precede activity i
  R_k is the availability level of resource type k over the project horizon
  A_t is the set of activities ongoing at time t
  r_{ik} is the number of units of resource type k required by activity i.

The objective of the aforementioned model is to minimize the completion time of the last activity in the project. Since the completion time of the last activity determines the project duration, the project duration is indirectly minimized. The first constraint set ensures that all predecessors of activity i are completed before activity i may start. The second constraint set ensures that resource allocation does not exceed resource availability. The general model may be modified or extended to consider other project parameters. Examples of other factors that may be incorporated into the scheduling formulation include cost, project deadline, activity contingency, mutual exclusivity of activities, activity crashing requirements, and activity subdivision.

An *objective function* is a mathematical representation of the goal of an organization. It is stated in terms of maximizing or minimizing some quantity of interest. In a project environment, the objective function may involve any of the following:

Minimize project duration.
Minimize project cost.
Minimize number of late jobs.
Minimize idle resource time.
Maximize project revenue.
Maximize net present worth.

LINEAR PROGRAMMING FORMULATION

Although many optimization models have emerged over the past decades, the linear programming (LP) formulation remains the seminal model from which many enhancements have been developed. If the formulation and framework of LP is understood, the whole concept of mathematical optimization becomes easier to accept.

Linear programming (LP) is a mathematical technique for maximizing or minimizing some quantity, such as profit, cost, or time, to complete a project. It is one of the most widely used quantitative techniques. It is a mathematical technique for finding the optimum solution to a linear objective function of two or more quantitative decision variables subject to a set of linear constraints. The technique is applicable to a wide range of decision-making problems. Its wide applicability is due to the fact that its formulation is not tied to any particular class of problems, as the CPM and PERT techniques are. Numerous research and application studies of LP are available in the literature.

The objective of a LP model is to optimize an objective function by finding values for a set of decision variable subject to a set of constraints.

The word *programming* in LP does not refer to computer programming, as some people think. Rather, it refers to choosing a *program of action*. The word *linear* refers to the *linear relationships* among the variables in the model. The characteristics of LP formulation are explained next.

Quantitative decision variables. A decision variable is a factor that can be manipulated by the decision maker. Examples are number of resource units assigned to a task, number of product types in a product mix, and number of units of a product to produce. Each decision variable must be defined numerically in some unit of measurement.

Linear objective function. The objective function relates to the measure of performance to be minimized or maximized. There is a linear relationship among the variables that make up the objective function. The coefficient of each variable in the objective function indicates its per-unit contribution (positive or negative) toward the value of the objective function.

Linear constraints. Every decision problem is subject to some specific limitations or constraints. The constraints specify the restrictions on how the decision maker may manipulate the decision variables. Examples of decision constraints are capacity limitations, maximum number of resource units available, demand and supply requirements, and number of work hours per day. The relationships among the variables in constraint must be expressed as linear functions represented as equations or inequalities.

Nonnegativity constraint. The nonnegativity constraint is common to all LP problems. This requires that all decision variables are restricted to nonnegative values.

The general procedure for using a LP model to solve a decision problem involves an LP formulation of the problem and a selection of a solution approach. The procedure is summarized as follows:

1. Determine the decision variables in the problem.
2. Determine the objective of the problem.

3. Formulate the objective function as an algebraic expression.
4. Determine the real-world restrictions on the problem scenario.
5. Write each of the restrictions as an algebraic constraint. Make sure that units match throughout the constraints. Otherwise, the terms cannot be added.
6. Select a solution approach. The *graphical method* and the *simplex technique* are the two most popular approaches. The graphical method is easy to apply when the LP model contains just two decision variables. Several commercial software packages are available for solving LP and related formulations.

An important aspect of using LP models is the interpretation of the results to make decisions. An LP solution that is optimal analytically may not be practical in a real-world decision scenario. The decision maker must incorporate their own subjective judgment when implementing LP solutions. Final decisions are often based on a combination of quantitative and qualitative factors. The examples presented in this chapter illustrate the application of optimization models to project planning and scheduling problems.

ACTIVITY PLANNING FORMULATION

Activity planning is a major function in any supply chain. LP can be used to determine the optimal allocation of time and resources to the activities in a project. Suppose a program planner is faced with the problem of planning a 5-day development program for a group of managers in a manufacturing organization. The program includes some combination of four activities: A seminar, laboratory work, case studies, and management games. It is estimated that each day spent on an activity will result in productivity improvement for the organization. The productivity improvement will generate annual cost savings as shown in Table 6.1. The program will last 5 days and there is no time lost between activities. In order to balance the program, the planner must make sure that not more than 3 days are spent on active or passive elements of the program. The active and passive percentages of each activity are also shown in the table. The company wishes to spend at least half a day on each of the four activities. A total budget of $1500 is available. The cost of each activity is shown in the tabulated data.

TABLE 6.1
Data for Activity Planning Problem

Activity	Cost Savings ($/year)	% Active	% Passive	Cost ($/day)
Seminar	3,200,000	10	90	400
Laboratory work	2,000,000	40	60	200
Case studies	400,000	100	0	75
Management games	2,000,000	60	40	100

The program planner must determine how many days to spend on each of the four activities. The following variables are defined for the problem:

x_1 represents number of days spent on a seminar
x_2 represents number of days of laboratory work
x_3 represents number of days for case studies
x_4 represents number of days with management games.

The objective is to maximize the estimated annual cost savings. That is,

$$\text{Maximize}: f = 3200x_1 + 2000x_2 + 400x_3 + 2000x_4$$

Subject to the following constraints:

1. The program lasts exactly 5 days.

$$x_1 + x_2 + x_3 + x_4 = 5$$

2. Not more than 3 days can be spent on active elements.

$$0.10x_1 + 0.40x_2 + x_3 + 0.60x_4 \leq 3$$

3. Not more than 3 days can be spent on passive elements.

$$0.90x_1 + 0.60x_2 + 0.40x_4 \leq 3$$

4. At least 0.5 day most be spent on each of the four activities.

$$x_1 \geq 0.50$$
$$x_2 \geq 0.50$$
$$x_3 \geq 0.50$$
$$x_4 \geq 0.50$$

5. The budget is limited to $1500.

$$400x_1 + 200x_2 + 75x_3 + 100x_4 \leq 1500$$

The complete LP model for the example is presented as follows.

$$\text{Maximize}: x_1 + x_2 + x_3 + x_4 = 5$$

The optimal solution to the problem is shown in Table 6.2. Most of the conference time must be allocated to the seminar (2.20 days).

TABLE 6.2
LP Solution to the Activity Planning Example

Activity	Cost Savings ($/year)	Number of Days	Annual Cost Savings ($)
Seminar	3,200,000	2.20	7,040,000
Laboratory work	2,000,000	0.50	1,000,000
Case studies	400,000	0.50	200,000
Management games	2,000,000	1.80	3,600,000
Total		5	11,840,000

The expected annual cost savings due to this activity is $7,040,000. That is, 2.20 days × $3,200,000/year/day. Management games are the second most important activity. A total of 1.8 days for management games will yield annual cost savings of $3,600,000. Fifty percent of the remaining time (0.5 day) should be devoted to laboratory work, which will result in annual cost savings of $1,000,000. Case studies also require half a day with a resulting annual savings of $200,000. The total annual savings, if the LP solution is implemented, is $11,840,000. Thus, an investment of $1,500 in management training for the personnel can generate annual savings of $11,840,000, a huge rate of return on investment!

RESOURCE COMBINATION FORMULATION

This example illustrates the use of LP for energy resource allocation. Suppose an industrial establishment uses energy for heating, cooling, and lighting. The required amount of energy is presently being obtained from conventional electric power and natural gas. In recent years, there have been frequent shortages of gas, and there is a pressing need to reduce the consumption of conventional electric power. The director of the energy management department is considering a solar energy system as an alternate source of energy. The objective is to find an optimal mix of three different sources of energy to meet the plant's energy requirements. The three energy sources are

Natural gas
Conventional electric power
Solar power.

It is required that the energy mix yield the lowest possible total annual cost of energy for the plant. Suppose a forecasting analysis indicates that the minimum kwh (kilowatt-hour) needed per year for heating, cooling, and lighting, are 1,800,000, 1,200,000, and 900,000 kwh, respectively. The solar energy system is expected to supply at least 1,075,000 kwh annually. The annual use of conventional electric power must be at least 1,900,000 kwh due to a prevailing contractual agreement for energy supply. The annual consumption of the contracted supply of gas must

TABLE 6.3
Energy Resource Combination Data

Energy Source	Supply (1000s kwh)	Savings (1000s $)	Unit Savings ($/kwh)	Unit Cost ($/kwh)
Solar Power	1,075	600	6	0.40
Electric Power	1,900	800	3	0.30
Natural Gas	950	375	2	0.20

TABLE 6.4
Tabulation of Data for LP Model

Energy Source	Heating	Type of Use Cooling	Lighting	Constraint
Solar Power	X_{11}	X_{12}	X_{13}	≥ 1,075K
Electric Power	X_{21}	X_{22}	X_{23}	≥ 1,900K
Natural Gas	X_{31}	X_{32}	X_{33}	≥ 950K
Constraint	≥ 1,800	≥ 1,200	≥ 900	

be at least 950,000 kwh. The cubic foot unit for natural gas has been converted to kwh (1 ft3 of gas = 0.3024).

The respective rates of $6, $3, and $2 per kwh are applicable to the three sources of energy. The minimum individual annual savings desired are $600,000 from solar power, $800,000 from conventional electric power, and $375,000 from natural gas. The savings are associated with the operating and maintenance costs. The energy cost per kwh is $0.30 for conventional electric power, $0.20 for natural gas, and $0.40 for solar power. The initial cost of the solar energy system has been spread over its useful life of 10 years with appropriate cost adjustments to obtain the rate per kwh. The problem data is summarized in Table 6.3. If we let x_{ij} be the kwh used from source i for purpose j, then we would have the data organized as shown in Table 6.4.

The optimization problem involves the minimization of the total cost function, Z. The mathematical formulation of the problem is presented as follows:

$$\text{Minimize}: Z = 0.4\sum_{j=1}^{3}x_{1j} + 0.3\sum_{j=1}^{3}x_{2j} + 0.2\sum_{j=1}^{3}x_{3j}$$

The solution to this example is presented in Table 6.5. The table shows that solar power should not be used for cooling and lighting if the lowest cost is to be realized. The use of conventional electric power should be spread over the three categories of use. The solution indicates that natural gas should be used for cooling purposes. In pragmatic terms, this LP solution may have to be modified before being implemented on the basis of the prevailing operating scenarios and the technical aspects of the units involved.

TABLE 6.5

LP Solution to the Resource Combination Example

Energy Source	Type of Use		
	Heating	Cooling	Lighting
Solar Power	1,075	0	0
Electric Power	750	250	900
Natural Gas	0	950	0

RESOURCE REQUIREMENT ANALYSIS

Activity–resource assignment combinations provide opportunities for finding the best allocation of resources to meet project goals within the prevailing constraints in the project environment (Badiru, 1993, 2019). Suppose a manufacturing project requires that a certain number of workers be assigned to a workstation. The workers produce identical units of the same product. The objective is to determine the number of workers to assign to the workstation in order to minimize the total production cost per shift. Each shift is 8 hours long. Each worker can be assigned a variable number of hours and/or variable production rates to work during a shift. Four different production rates are possible: *Slow rate, normal rate, fast rate*, and *high-pressure rate*. Each worker is capable of working at any of the production rates during a shift. The total number of work hours available per shift is determined by multiplying the number of workers assigned by the 8 hours available in a shift.

There are variable costs and percent defective associated with each production rate. The variable cost and the percent defective increase as the production rate increases. At least 450 units of the product must be produced during each shift. It is assumed that the workers' performance levels are identical. The production rates (r_i), the respective costs (c_i), and percent defective (d_i), are presented as follows:

Operating Rate 1 (Slow)

$$r_1 = 10 \text{ units/h}$$
$$c_1 = \$5/h$$
$$d_1 = 5\%$$

Operating Rate 2 (Normal)

$$r_2 = 18 \text{ units/h}$$
$$c_2 = \$10/h$$
$$d_2 = 5\%$$

Operating Rate 3 (Fast)

$$r_3 = 30 \text{ units/h}$$
$$c_3 = \$15/\text{h}$$
$$d_3 = 12\%$$

Operating Rate 4 (High Pressure)

$$r_4 = 40 \text{ units/h}$$
$$c_4 = \$25/\text{h}$$
$$d_4 = 15\%$$

Let xi represent the number of hours worked at production rate i.
Let n represent the number of workers assigned.
Let u_i represent the number of good units produced at operation rate i.

$$u_1 = \left(10 \text{ units/h}\right) \cdot \left(1 - 0.05\right) = 9.50 \text{ units/h}$$
$$u_2 = \left(18 \text{ units/h}\right) \cdot \left(1 - 0.08\right) = 16.56 \text{ units/h}$$
$$u_3 = \left(30 \text{ units/h}\right) \cdot \left(1 - 0.12\right) = 24.40 \text{ units/h}$$
$$u_4 = \left(40 \text{ units/h}\right) \cdot \left(1 - 0.15\right) = 34.00 \text{ units/h}$$

$$\text{Minimize}: z = 5x_1 + 10x_2 + 15x_3 + 25x_4$$

$$x_1 + x_2 + x_3 + x_4 \leq 8n$$
$$\text{Subject to}: \quad 9.50x_1 + 16.56x_2 + 25.40x_3 + 34.00x_4 \leq 450$$
$$x_1, x_2, x_3, x_4 \geq 0$$

The solution will be obtained by solving the LP model for different values of n. A plot of the minimum costs versus values of n can then be used to determine the optimum assignment policy. The complete solution can be explore further by interested readers.

INTEGER PROGRAMMING FORMULATION

Integer programming is a restricted model of LP that permits only solutions with integer values of the decision variables. Suppose we are interested in minimizing the project completion time while observing resource limitations and job precedence relationships. The basic assumption is that once a job starts, it has to be completed without interruption. We can construct several different integer programming models for the problem, all of which will give the same optimal solution but whose execution times will differ considerably. An efficient integer programming model for the problem should use as few integer variables as possible.

Define variables as

$$x_{ij} : 1 \text{ if job } i \text{ starts in period } j; \quad 0, \text{ otherwise}$$

$$t_p : \text{completion time of the project}$$

Only x_{ij}'s are restricted integers. For each job i one can determine the early start and latest start times, ES_i and LS_i, respectively. Therefore, assuming that there are n jobs in the project, we have $1 \leq i \leq n$ for each i, then we have $ES_i \leq j \leq LS_i$. Let t_i denote the duration of job i. Resource availability constraints can be smartly handled by defining a vector V_{ij} which has 0's everywhere except positions $j, j+1,..., j+ti-1$, where it has 1's. It indicates the time period where job i uses the resource assuming that $x_{ij} = 1$. Let r_i and R_j be the resource required by job i and the resource available on day j, respectively. Let R be a row vector containing R_j.

Then, the integer programming model for the scheduling problem with limited resource can be defined as

Minimize t_p

Subject to:

$$\sum_{j=ES_i}^{LS_i} x_{ij} = 1 \quad \forall i = 1,..., n$$

$$-\sum_{j=ES_i}^{LS_i} jx_{ij} + \sum_{j=ES_k}^{LS_k} jx_{kj} \leq t_i \quad \forall k \in S(i)$$

$$\forall i = 1,..., n$$

$$t_p - \sum_{j=ES_i}^{LS_i} jx_{ij} \geq t_i - 1 \quad \forall i \text{ with } S(i) = \phi$$

$$\sum_{i=1}^{n} \sum_{j=ES_i}^{LS_i} x_{ij} r_i V_{ij} \leq R$$

$$x_{ij} = 0, 1 \quad \forall i = ES_i,..., LS_i$$

$$\forall i = 1,..., n$$

where $S(i)$ is the set of immediate successor jobs of job i.

The first equation above indicates that each job must start on the same day. The second equation makes sure that a job cannot start until all of the predecessor jobs are completed. The third equation determines the project completion time t_p. The project is completed after all the jobs without any successors are completed. The last set of equations makes sure that daily resource requirements are met. The indicator variable x_{ij} is restricted to the values 0 and 1.

The previous integer programming model can be solved using LINDO computer code and declaring x_{ij}'s as binary variables. The code uses the branch and bound method of integer programming to solve the problem. Readers interested in the full details of the optimization modeling, examples, and solutions may refer to Badiru and Pulat (1995).

To get the best out of any system, we must think in terms of the best combination of goals, constraints, and resources. Only a mathematical representation and an optimization technique can handle the enormous combination of factors. Thus, the optimization techniques and formulations presented in this chapter offer a pathway to getting the most out of a supply chain, particularly in a case of global operational disruptions.

REFERENCES

Badiru, A. B. (1993). Activity resource assignments using critical resource diagramming. *Project Management Journal, 14*(3), 15–21.

Badiru, A. B. (2019). *Project management: Systems, principles, and applications* (2nd ed.). Taylor & Francis/CRC Press.

Badiru, A. B., & Pulat, P. S. (1995). *Comprehensive project management: Integrating optimization models, management principles, and computers* (pp. 162–209). Prentice Hall.

Montazer, M. A., & Shahbaz, A. R. (2023, May 18). *Role of industrial engineering and operations research (IEOR) in improvement of United States postal service (USPS) during 1600 to 1950.* Paper presented at the 2023 Annual Conference of IISE (Institute of Industrial and Systems Engineers), New Orleans, LA.

7 Supply Chain Forecasting

FORECASTING FOR THE SUPPLY CHAIN

The models, tools, techniques, and concepts for forecasting found in general business and industry enterprises are applicable to the general postal service industry. Any supply chain is subject to dynamic changes in production, shipment, and delivery processes. For the supply to be adaptive, responsive, and resilient to the changes, either expected or unexpected, a combination of analytical, qualitative, and computer techniques must be employed. Good forecasting is the basis for achieving a responsive supply chain. Hogan et al. (2020), Badiru et al. (1993), Badiru (2019), and references therein present tools and techniques pertinent for forecasting in the supply chain environment.

Managing complex supply chains effectively calls for good information, which can be provided by forecasting and inventory control. Forecasting is not just for marketing and production planning purposes. This chapter presents techniques of forecasting and inventory management as a part of the overall quantitative techniques for supply chain planning, design, and control. Several analytical tools are essential for analyzing project systems. These are relevant if we formulate a supply chain as a project system as discussed in Chapter 1. Prior to proceeding to the project management phase, a good understanding of the enterprise system, within which the supply chain resides, is indispensable for getting a successful output.

Forecasting should be an important part of overall supply strategy. Effective prediction provides information needed to make good enterprise-wide decisions. Several techniques are available for forecasting. Regression, time series analysis, computer simulation, and artificial neural networks are common examples of forecasting techniques. There are two basic types of forecasting: *Intrinsic forecasting* and *extrinsic forecasting*.

Intrinsic forecasting assumes that historical data adequately describe the problem scenario to be forecasted. Forecasting models based on historical data require extrapolation to generate estimates for the future. The requirements of intrinsic forecasting are:

- Collect historical data.
- Develop quantitative forecasting model based on the data collected.
- Generate forecasts recursively for the future.
- Revise the forecasts as new pieces of data become available.

Extrinsic forecasting assumes that the forecasts to be generated are correlated to some other external factors such that the forecasts of the external factors provide reliable forecasts for the current problem. For example, the demand for a new product

 DOI: 10.1201/9781032620701-7

may be based on forecasts of household incomes. Before any forecasting system is implemented, a complete analysis of the data required must be performed. This is useful for setting activity times and task allocation strategies.

DATA MEASUREMENT SCALES FOR FORECASTING

Forecasting requires data collection, measurement, and analysis. In the supply chain, the analyst will encounter different types of measurement scales depending on the particular items involved. Data may need to be collected on shipment schedules, costs, performance levels, problems, and so on. The different types of data measurement scales that are applicable are presented below.

Nominal scale of measurement: A *nominal scale* is the lowest level of measurement scales. It classifies items into categories. The categories are mutually exclusive and collectively exhaustive. That is, the categories do not overlap and they cover all possible categories of the characteristics being observed. For example, in the analysis of the critical path in a project network, each job is classified as either critical or not critical. Gender, type of industry, job classification, and color are some examples of measurements on a nominal scale.

Ordinal scale of measurement: An *ordinal scale* is distinguished from a nominal scale by the property of order among the categories. An example is the process of prioritizing project tasks for resource allocation. We know that first is above second, but we do not know how far above. Similarly, we know that better is preferred to good, but we do not know by how much. In quality control, the ABC classification of items based on the Pareto distribution is an example of a measurement on an ordinal scale.

Interval scale of measurement: An *interval scale* is distinguished from an ordinal scale by having equal intervals between the units of measure. The assignment of priority ratings to project objectives on a scale of 0–10 is an example of a measurement on an interval scale. Even though an objective may have a priority rating of 0, it does not mean that the objective has absolutely no significance to the project team. Similarly, the scoring of 0 on an examination does not imply that a student knows absolutely nothing about the materials covered by the examination. Temperature is a good example of an item that is measured on an interval scale. Even though there is a zero point on the temperature scale, it is an arbitrary relative measure. Other examples of interval scales are IQ measurements and aptitude ratings.

Ratio scale of measurement: A *ratio scale* has the same properties of an interval scale but with a true zero point. For example, an estimate of a zero time unit for the duration of a task is a ratio scale measurement. Other examples of items measured on a ratio scale are cost, time, volume, length, height, weight, and inventory level. Many of the items measured in a project management environment will be on a ratio scale.

Another important aspect of data analysis for project control involves the classification scheme used. Most projects will have both *quantitative* and *qualitative* data. Quantitative data require that we describe the characteristics of the items being studied numerically. Qualitative data, on the other hand, are associated with object attributes that are not measured numerically. Most items measured on the nominal and ordinal scales will normally be classified into the qualitative data category while those measured on the interval and ratio scales will normally be classified into the quantitative data category.

The implication for project control is that qualitative data can lead to bias in the control mechanism because qualitative data are subject to the personal views and interpretations of the person using the data. Whenever possible, data for project control should be based on quantitative measurements.

There is a class of project data referred to as *transient data*. This is defined as a volatile set of data that is used for one-time decision making and is not then needed again. An example may be the number of operators that show up at a job site on a given day. Unless there is some correlation between the day-to-day attendance records of operators, this piece of information will have relevance only for that given day. The project manager can make his decision for that day on the basis of that day's attendance record. Transient data need not be stored in a permanent database unless it may be needed for future analysis or uses (e.g., forecasting, incentive programs, and performance review).

Recurring data refer to data that are encountered frequently enough to necessitate storage on a permanent basis. An example is a file containing contract due dates. This file will need to be kept at least through the project life cycle. Recurring data may be further categorized into *static data* and *dynamic data*. Recurring data that are static will retain their original parameters and values each time they are retrieved and used. Recurring data that are dynamic have the potential for taking on different parameters and values each time they are retrieved and used. Storage and retrieval considerations for project control should address the following questions:

1. What is the origin of the data?
2. How long will the data be maintained?
3. Who needs access to the data?
4. What will the data be used for?
5. How often will the data be needed?
6. Are the data for lookup purposes only (i.e., no printouts)?
7. Are the data for reporting purposes (i.e., generate reports)?
8. In what format are the data needed?
9. How fast will the data need to be retrieved?
10. What security measures are needed for the data?

DATA DETERMINATION AND COLLECTION

It is essential to determine what data to collect for project control purposes. Data collection and analysis are basic components of generating information for project control. The requirements for data collection are discussed next.

Choosing the Data

This involves selecting data on the basis of their relevance and the level of likelihood that they will be needed for future decisions and whether or not they contribute to making the decision better. The intended users of the data should also be identified.

Collecting the Data

This identifies a suitable method of collecting the data as well as the source from which the data will be collected. The collection method will depend on the particular operation being addressed. The common methods include manual tabulation, direct keyboard entry, optical character reader, magnetic coding, electronic scanner, and more recently, voice command. An input control may be used to confirm the accuracy of collected data. Examples of items to control when collecting data include the following.

Relevance Check

This checks if the data are relevant to the prevailing problem. For example, data collected on personnel productivity may not be relevant for a decision involving marketing strategies.

Limit Check

This checks to ensure that the data are within known or acceptable limits. For example, an employee overtime claim amounting to over 80 hours/week for several weeks in a row is an indication of a record well beyond ordinary limits.

Critical Value

This identifies a boundary point for data values. Values below or above a critical value fall in different data categories. For example, the lower specification limit for a given characteristic of a product is a critical value that determines whether or not the product meets quality requirements.

Coding the Data

This refers to the technique used in representing data in a form useful for generating information. This should be done in a compact and yet meaningful format. The performance of information systems can be greatly improved if effective data formats and coding are designed into the system right from the beginning.

Processing the Data

Data processing is the manipulation of data to generate useful information. Different types of information may be generated from a given data set depending on how it is processed. The processing method should consider how the information will be used, who will be using it, and what caliber of system response time is desired. If possible, processing controls should be used.

Control Total

Check for the completeness of the processing by comparing accumulated results to a known total. An example of this is the comparison of machine throughput to a standard production level or the comparison of cumulative project budget depletion to a cost accounting standard.

Consistency Check

Check if the processing is producing the same results for similar data. For example, an electronic inspection device that suddenly shows a measurement that is 10 times higher than the norm warrants an investigation of both the input and the processing mechanisms.

Scales of Measurement

For numeric scales, specify units of measurement, increments, the zero point on the measurement scale, and the range of values.

Using the Information

Using information involves people. Computers can collect data, manipulate data, and generate information, but the ultimate decision rests with people, and decision making starts when information becomes available. Intuition, experience, training, interest, and ethics are just a few of the factors that determine how people use information. The same piece of information that is positively used to further the progress of a project in one instance may also be used negatively in another instance. To assure that data and information are used appropriately, computer-based security measures can be built into the information system.

Project data may be obtained from several sources. Some potential sources are

- Formal reports
- Interviews and surveys
- Regular project meetings
- Personnel time cards or work schedules.

The timing of data is also very important for project control purposes. The contents, level of detail, and frequency of data can affect the control process. An important aspect of project management is the determination of the data required to generate the information needed for project control. The function of keeping track of the vast quantity of rapidly changing and interrelated data about project attributes can be very complicated. The major steps involved in data analysis for project control are

- Data collection
- Data analysis and presentation
- Decision making
- Implementation of action.

Data are processed to generate information. Information is analyzed by the decision maker to make the required decisions. Good decisions are based on timely and relevant information, which in turn is based on reliable data. Data analysis for project control may involve the following functions:

- Organizing and printing computer-generated information in a form usable by managers;
- Integrating different hardware and software systems to communicate in the same project environment;
- Incorporating new technologies such as expert systems into data analysis;

- Using graphics and other presentation techniques to convey project information.

Proper data management will prevent misuse, misinterpretation, or mishandling. Data are needed at every stage in the life cycle of a project from the problem identification stage through the project phase-out stage. The various items for which data may be needed are project specifications, feasibility study, resource availability, staff size, schedule, project status, performance data, and phase-out plan. The documentation of data requirements should cover the following:

Data summary. A data summary is a general summary of the information and decision for which the data are required as well as the form in which the data should be prepared. The summary indicates the impact of the data requirements on the organizational goals.

Data processing environment. The processing environment identifies the project for which the data are required, the user personnel, and the computer system to be used in processing the data. It refers to the project request or authorization and relationship to other projects and specifies the expected data communication needs and mode of transmission.

Data policies and procedures. Data handling policies and procedures describe policies governing data handling, storage, and modification and the specific procedures for implementing changes to the data. Additionally, they provide instructions for data collection and organization.

Static data. A static data description describes that portion of the data that are used mainly for reference purposes and it is rarely updated.

Dynamic data. A dynamic data description describes that portion of the data that are frequently updated based on the prevailing circumstances in the organization.

Data frequency. The frequency of data update specifies the expected frequency of data change for the dynamic portion of the data (e.g., quarterly). This data change frequency should be described in relation to the frequency of processing.

Data constraints. Data constraints refer to the limitations on the data requirements. Constraints may be procedural (e.g., based on corporate policy), technical (e.g., based on computer limitations), or imposed (e.g., based on project goals).

Data compatibility. Data compatibility analysis involves ensuring that data collected for project control needs will be compatible with future needs.

Data contingency. A data contingency plan concerns data security measures in case of accidental or deliberate damage or sabotage affecting hardware, software, or personnel.

FORECASTING BASED ON AVERAGES

The most common forecasting techniques are based on averages. Sophisticated quantitative forecasting models can be formulated from basic average formulas. The traditional methods of averages are presented as follows.

SIMPLE AVERAGE FORECAST

In this method, the forecast for the next period is computed as the arithmetic average of the preceding data points. This is often referred to as average to date. That is,

$$f_{n+1} = \frac{\sum_{t=1}^{n} d}{n}$$

where

f_{n+1} is the forecast for period $n+1$
d is the data element for the period in question
n is the number of preceding periods for which data are available

PERIOD MOVING AVERAGE FORECAST

In this method, the forecast for the next period is based only on the most recent data values. Each time a new value is included, the oldest value is dropped. Thus, the average is always computed from a fixed number of values. This is represented as

$$f_{n+1} = \frac{\sum_{t=n-T+1}^{n} d_t}{T}$$
$$= \frac{d_{n-T+1} + d_{n-T+2} + \cdots + d_{n-1} + d_n}{T}$$

where

f_{n+1} is the forecast for period $n+1$
d_t is the datum for period t
T is the number of preceding periods included in the moving average calculation
n is the current period at which forecast of f_{n+1} is calculated

The moving average technique is an after-the-fact approach. Since T data points are needed to generate a forecast, we cannot generate forecasts for the first $T-1$ periods. But this shortcoming is quickly overcome as the number of data points available becomes large.

WEIGHTED AVERAGE FORECAST

The weighted average forecast method assumes that some data points might be more significant that others in generating future forecasts. For example, the most recent data points may weigh more than very old data points in the calculation of future estimates. This is expressed as

$$f_{n+1} = \frac{\sum_{t=1}^{n} w_i d_t}{\sum_{t=1}^{n} w_t}$$
$$= \frac{w_1 d_1 + w_2 d_2 + \cdots + w_n d_n}{w_1 + w_2 + \cdots + w_n}$$

where

f_{n+1} is the weighted average forecast for period $n+1$

d_t is the datum for period t

T is the additional notation representing the planning horizon for the forecast problem

n is the current period at which forecast of f_{n+1} is calculated

w_t is the weight of data point t

The w_t's are the respective weights of the data points such that

$$\sum_{t=1}^{n} w_t = 1.0$$

WEIGHTED T-PERIOD MOVING AVERAGE FORECAST

In this technique, the forecast for the next period is computed as the weighted average of past data points over the last T time periods. That is,

$$f_{n+1} = w_1 d_n + w_2 d_{n-1} + \cdots + w_T d_{n-T+1}$$

where w_i's are the respective weights of the data points such that

$$\sum_{i=1}^{n} w_i = 1.0$$

EXPONENTIAL SMOOTHING FORECAST

This is a special case of weighted moving average forecast. The forecast for the next period is computed as the weighted average of the immediate past data point and the forecast of the previous period. In order words, the previous forecast is adjusted based on the deviation (forecast error) of that forecast from the actual data. That is,

$$f_{n+1} = \alpha d_n + (1-\alpha) f_n$$
$$= f_n + \alpha (d_n - f_n)$$

where

f_{n+1} is the exponentially weighted average forecast for period $n+1$

d_n is the datum for period n

f_n is the forecast for period n

α is the smoothing factor (real number between 0 and 1)

A low smoothing factor gives a high degree of smoothing, while a high value causes the forecast to closely match actual data.

REGRESSION ANALYSIS

The primary function of regression analysis is to develop a model which expresses the relationship between a dependent variable and one or more independent variables.

It is sometimes called line fitting or curve fitting. Regression analysis is an important statistical tool that can be applied to many prediction and forecasting problems in the project environment. The utility of a regression model is often tested by analysis of variance (ANOVA), which is a technique for breaking down the variance in a statistical sample into components that can be attributed to each factor affecting that sample. One major purpose of ANOVA is testing of the model. Model testing is important because of the serious consequences of erroneously concluding that a regression model is good when, in fact, it has little or no significance to the data. Model inadequacy often implies an error in the assumed relationships between the variables, poor data, or both. A validated regression model can be used for the following purposes:

1. Prediction/forecasting
2. Description
3. Control.

DESCRIPTION OF REGRESSION RELATIONSHIP

Sometimes, the desired result from a regression analysis is an equation describing the best fit to the data under investigation. The "least squares" line drawn through the data is the line of best fit. This line may be linear or curvilinear depending on the dispersion of the data. The linear situation exists in those cases where the slope of the regression equation is a constant. The nonconstant slope indicates curvilinear relationships. A plot of the data, called scatter plot, will usually indicate whether a linear or nonlinear model will be appropriate. The major problem with the nonlinear relationship is the necessity of assuming a functional relationship before accurately developing the model. Example of regression models (simple linear, multiple, and nonlinear) are presented as follows:

$$Y = \beta_0 + \beta_1 x + \varepsilon$$
$$Y = \beta_0 + \beta_1 x_1 + \beta_2 x_2 + \varepsilon$$
$$Y = \beta_0 + \beta_1 x_1^{\alpha 1} + \beta_2 x_2^{\alpha 2} + \varepsilon$$
$$Y = \beta_0 + \beta_1 x_1^{\alpha 1} + \beta_2 x_2^{\alpha 2} + \beta_{12} x_1^{\alpha 3} x_2^{\alpha 4} + \varepsilon$$

where
Y is the dependent variable
x_i's are the independent variables
β_i's are the model parameters
ε is the error term

The error terms are assumed to be independent and identically distributed normal random variables with mean of zero and variance of σ^2.

PREDICTION

Another major use of regression analysis is prediction or forecasting. Prediction can be of two basic types: Interpolation and extrapolation. Interpolation predicts values of the dependent variable over the range of the independent variable or variables. Extrapolation involves predictions outside the range of the independent variables. Extrapolation carries a risk in the sense that projections are made over a data range that is not included in the development of the regression model. There is some level of uncertainty about the nature of the relationships that may exist outside the study range. Interpolation can also create a problem when the values of the independent variables are widely spaced.

CONTROL

Extreme care is needed in using regression for control. The difficulty lies in the assumption of a functional relationship when in fact none exists. Suppose, for example, that regression shows a relationship between chemical content in a product and noise level in the room. Suppose further that the real reason for this relationship is that the noise level increases as the machine speed increases and higher machine speed produces higher chemical content. It would be erroneous to assume a functional relationship between the noise level in the room and the chemical content in the product. If this relationship does exist, then changes in the noise level could control chemical content. In this case, the real functional relationship exists between machine speed and chemical content. It is often difficult to prove functional relationships outside a laboratory environment because many extraneous and intractable factors may have an influence on the dependent variable. A simple example of the use of functional relationship for control can be seen in the following familiar equation of electrical circuits:

$$I = \frac{V}{R}$$

where
    V is voltage
    I is electrical current
    R is the resistance

The current can be controlled by changes in either the voltage or the resistance or both. This particular equation, which has been experimentally validated, can be used as a control device.

PROCEDURE FOR REGRESSION ANALYSIS

Problem definition: Failure to properly define the scope of the problem could result in useless conclusions. Time can be saved throughout all phases of a regression study by knowing, as precisely as possible, the purpose of the required model. A proper definition of the problem will facilitate the selection of the appropriate variables to include in the study.

Selection of variables: Two very important factors in the selection of variables are ease of data collection and expense of data collection. Ease of data collection deals with the accessibility and the desired form of data. We must first determine if the data can be collected and, if so, how difficult the process will be. The economic question is of prime importance. How expensive will the data be to collect and compile into a useable form? If the expense cannot be justified, then the variable under consideration may necessarily be omitted from the selection process.

Test of significance of regression: After the selection and compilation of all possible relevant variables, the next step is a test for the significance of regression. The test should help avoid wasted effort on the use of an invalid model. The test for the significance of regression is a test to see if at least one of the variable coefficient(s) in the regression equation is statistically different from zero. A test indicating that none of the coefficients is significantly different from zero implies that the best approximation of the data is a straight line through the data at the average value of the dependent variable regardless of the values of the independent variables. The significance level of the data is an indication of the probability of erroneously assuming model validity.

COEFFICIENT OF DETERMINATION

The coefficient of multiple determination, denoted by R^2, is used to judge the effectiveness of regression models containing multiple variables (multiple regression model). It indicates the proportion of the variation in the dependent variable explained by the model. The coefficient of multiple determination is defined as

$$R^2 = \frac{SSR}{SST}$$

$$= 1 - \frac{SSE}{SST}$$

where
  SSR represents the sum of squares due to the regression model
  SST represents the sum of squares total
  SSE represents the sum of squares due to error

R^2 measures the proportionate reduction of total variation in the dependent variable accounted for by a specific set of independent variables. The coefficient of multiple determination, R^2, reduces to the *coefficient of simple determination, r^2,* when there is only one independent variable in the regression model. R^2 is equal to 0 when all the coefficients, b_k, in the model are zero. That is, no regression fit at all. R^2 is equal to one when all data points fall directly on the fitted response surface. Thus, we have

$$0.0 \le R^2 \le 1.0$$

The following points should be noted about regression modeling:

1. A large R^2 does not necessarily imply that the fitted model is a useful one. For example, observations may have been taken at only a few levels

of the independent variables. In such a case, the fitted model may not be useful because most predictions would require extrapolation outside the region of observations. For example, for only two data points, the regression line passes perfectly through the two points and the R^2 value will be one. In that case, despite the high R^2, there will be no useful prediction capability.

2. Adding more independent variables to a regression model can only increase R^2 and never reduce it. This is because the error sum of squares (SSE) can never become larger with more independent variables and the total sum of squares (SST) is always the same for a given set of responses.

3. Regression models developed under conditions where the number of data points is roughly equal to the number of variables will yield high values of R^2 even though the model may not be useful. For example, for only two data points, the regression line will pass perfectly through the two points and r^2 will be one. Even though r^2 is one, there will be no useful prediction.

The strategy for using R^2 to evaluate regression models should not entirely focus on maximizing the R^2 value. Rather, the intent should be to find the point where adding more independent variables is not worthwhile in terms of the overall effectiveness of the regression model. For most practical situations, R^2 values greater than 0.62 are considered acceptable.

Since R^2 can often be made larger by including a large number of independent variables, it is sometimes suggested that a modified measure be used which recognizes the number of independent variables in the model. This modified measure is referred to as adjusted coefficient of multiple determination, R_a^2. It is defined mathematically as

$$R_a^2 = 1 - \left(\frac{n-1}{n-p} \right) \frac{SSE}{SST}$$

where
n is the number of observations used to fit the model
p is the number of coefficients in the model (including the constant term)
$p-1$ is the number of independent variables in the model
R_a^2 may actually become smaller when another independent variable is introduced into the model. This is because the decrease in SSE may be more than offset by the loss of a degree of freedom in the denominator, $n-p$.

The *coefficient of multiple correlation* is defined as the positive square root of R^2. That is,

$$R = \sqrt{R^2}$$

Thus, the higher the value of R^2, the higher the correlation in the fitted model.

RESIDUAL ANALYSIS

A residual is the difference between the predicted value computed from the fitted model and the actual value from the data. The ith residual is defined as

$$e_i = Y_i - \hat{Y}_i$$

where
    Y_i is the actual value
    $Y\hat{}_i$ is the predicted value

The sum of squares of errors, SSE, and the mean square error, MSE, are computed as

$$SSE = \sum_i e_i^2$$

$$\sigma^2 \approx \frac{\sum_i e_i^2}{n-2} = MSE$$

where n is the number of data points. A plot of residuals versus predicted values of the dependent variable can be quite revealing. The plot for a good regression model will have a random pattern. A noticeable trend in the residual pattern indicates a problem with the model. Some possible reasons for an invalid regression model are as follows:

    Insufficient data
    Important factors not included in model
    Inconsistency in data
    No functional relationship exists.

Graphical analysis of residuals is important for assessing the appropriateness of regression models. The different possible residual patterns are presented in Figure 7.1.

    When we plot the residuals versus the independent variable, the result should appear ideally as shown in the first plot. The second plot shows a residual pattern indicating nonlinearity of the regression function. The third plot shows a pattern suggesting nonconstant variance (i.e., variation in σ^2). The fourth plot presents a residual pattern implying interdependence of the error terms The fifth plot shows a pattern depicting the presence of outliers. The sixth and last plot represents a pattern suggesting omission of independent variables.

TIME SERIES ANALYSIS

Time series analysis is a technique that attempts to predict the future by using historical data. The basic principle of time series analysis is that the sequence of observations is based on jointly distributed random variables. The time series observations denoted by $Z_1, Z_2, ..., Z_T$ are assumed to be drawn from some joint probability density function (pdf) of the form

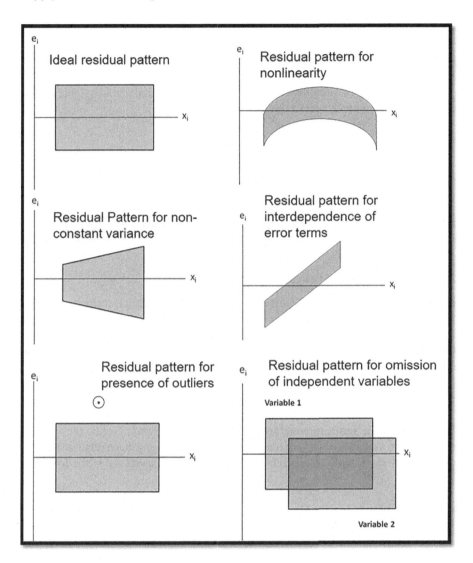

FIGURE 7.1 Six residual patterns.

$$f_{1,\ldots,T}\left(Z_{1,\ldots,T}\right)$$

The objective of time series analysis is to use the joint density to make probability inferences about future observations. The concept of stationarity implies that the distribution of the time series is invariant with regard to any time displacement. That is,

$$f\left(Z_t,\ldots,Z_{t+k}\right)=f\left(Z_{t+m},\ldots,Z_{t+m+k}\right)$$

where
  t is any point in time
  k and m are any pair of positive integers

A stationary time series process has a constant variance and remains stable around a constant mean with respect to time reference. Thus,

$$E(Z_t) = E(Z_{t+m})$$
$$V(Z_t) = V(Z_{t+m})$$
$$Cov(Z_t, Z_{t+1}) = Cov(Z_{t+k}, Z_{t+k+1})$$

Nonstationarity in a time series may be recognized in a plot of the series. A widely scattered plot with no tendency for a particular value is an indication of nonstationarity.

STATIONARITY AND DATA TRANSFORMATION

In some cases where nonstationarity exists, some form of data transformation may be used to achieve stationarity. For most time series data, the usual transformation tried is differencing. Differencing involves the creation of a new series by taking differences between successive periods of the original series. For example, first regular differences are obtained by

$$w_t = Z_t - Z_{t-1}$$

To develop a time series forecasting model, it is necessary to describe the relationship between a current observation and previous observations. Such relationships are described by the sample autocorrelation function defined as shown in the following formula:

$$r_j = \frac{\sum_{t=1}^{T-j}(Z_t - \bar{Z})(Z_{t+j} - \bar{Z})}{\sum_{t=1}^{T}(Z_t - \bar{Z})^2}, \quad j = 0, 1, ..., T-1$$

where
  T is the number of observations
  Z_t is the observation for time t
  $Z-$ is the sample mean of the series
  j is the number of periods separating pairs of observations
  r_j is the sample estimate of the theoretical correlation coefficient
  The coefficient of correlation between two variables Y_1 and Y_2 is defined as

$$\rho_{12} = \frac{\sigma_{12}}{\sigma_1 \sigma_2}$$

where

σ_1 and σ_2 are the standard deviations of Y_1 and Y_2, respectively

σ_{12} is the covariance between Y_1 and Y_2

The standard deviations are the positive square roots of the variances defined as

$$\sigma_1^2 = E\left[(^{Y_1}\right]$$
$$\sigma_2^2 = E\left[(^{Y_2}\right]$$

The covariance, σ_{12}, is defined as

$$\sigma_{12} = E\left[(Y_1 - \mu_1)(Y_2 - \mu_2)\right]$$

which will be zero if Y_1 and Y_2 are independent. Thus, when $\sigma_{12} = 0$, we also have $\rho_{12} = 0$ If Y_1 and Y_2 are positively related, then σ_{12} and ρ_{12} are both positive. If Y_1 and Y_2 are negatively related, then σ_{12} and ρ_{12} are both negative. The correlation coefficient is a real number between -1 and $+1$:

$$1.0 \le \rho_j \le +1.0$$

Time series modeling procedure involves the development of a discrete linear stochastic process in which each observation, Z_t, may be expressed as

$$Z_t = \mu + \mu_t + \Psi_1 u_{t-1} + \Psi_2 u_{t-2} + \cdots$$

where

μ is the mean of the process

Ψ_i's are model parameters, which are functions of the autocorrelations

Note that this is an infinite sum indicating that the current observation at time t can be expressed in terms of all previous observations from the past. In a practical sense, some of the coefficients will be zero after some finite point q in the past. The u_t's form the sequence of independently and identically distributed random disturbances with mean zero and variance sigma sub u^2. The expected value of the series is obtained by

$$E(Z_t) = \mu + E(u_t + \Psi_1 u_{t-1} + \Psi_2 u_{t-2} + \cdots)$$
$$= \mu + E(u_t)[1 + \Psi_1 + \Psi_2 + \cdots]$$

Stationarity of the time series requires that the expected value be stable. That is, the infinite sum of the coefficients should be convergent:

$$\sum_{i=0}^{\infty} \Psi_i = c$$

where
$$\Psi_0 = 1$$
c is a constant

Sample estimates of the variances and covariances are obtained by

$$c_j = \frac{1}{T} \sum_{t=1}^{T-j} (Z_t - \bar{Z})(Z_{t+j} - \bar{Z}), \quad j = 0,1,2,\ldots$$

The theoretical autocorrelations are obtained by dividing each of the autocovariances, γ_j, by γ_0. Thus, we have

$$\rho_j = \frac{\gamma_j}{\gamma_0}, \quad j = 0,1,2,\ldots$$

and the sample autocorrelations is obtained by

$$r_j = \frac{c_j}{c_0}, \quad j = 0,1,2,\ldots$$

Moving Average Processes

If it can be assumed that Ψ_i for some $i > q$, where q is an integer, then our time series model can be represented as

$$z_t = \mu + u_t + \Psi_1 u_{t-1} + \Psi_2 u_{t-2} + \cdots + \Psi_q u_{t-q}$$

which is referred to as a moving-average process of order q, usually denoted as $MA(q)$. For notational convenience, we will denote the truncated series as presented in the following:

$$Z_t = \mu + u_t - \theta_1 u_{t-1} - \theta_2 u_{t-2} - \cdots - \theta_q u_{t-q}$$

where $\theta_0 = 1$. Any $MA(q)$ process is stationary since the condition of convergence for the Ψ_is becomes

$$(1 + \Psi_1 + \Psi_2 + \cdots) = (1 - \theta_1 - \theta_2 - \cdots - \theta_q)$$

$$= 1 - \sum_{i=0}^{q} \theta_i$$

which converges since q is finite. The variance of the process now reduces to

$$\gamma_0 = \sigma_u^2 \sum_{i=0}^{q} \theta_i$$

We now have the autocovariances and autocorrelations defined, respectively, as shown in the following:

$$\gamma_j = \sigma_u^2 \left(-\theta_j + \theta_1\theta_{j+1} + \cdots + \theta_{q-j}\theta_q \right), \quad j = 1,\ldots,q$$

where $\gamma_j = 0$ for $j > q$

$$\rho_j = \frac{\left(-\theta_j + \theta_1\theta_{j+1} + \cdots + \theta_{q-j}\theta_q \right)}{\left(1 + \theta_1^2 + \cdots + \theta_q^2 \right)}, \quad j = 1,\ldots,q$$

where $\rho_j = 0$ for $j > q$.

Autoregressive Processes

In the preceding section, the time series, Z_t, is expressed in terms of the current disturbance, ut, and past disturbances, $ut - i$. An alternative is to express Z_t, in terms of the current and past observations, Z_{t-i}. This is achieved by rewriting the time series expression as

$$u_t = Z_t - \mu - \Psi_1 u_{t-1} - \Psi_2 u_{t-2} - \cdots$$
$$u_{t-1} = Z_{t-1} - \mu - \Psi_1 u_{t-2} - \Psi_2 u_{t-3} - \cdots$$
$$u_{t-2} = Z_{t-2} - \mu - \Psi_1 u_{t-3} - \Psi_2 u_{t-4} - \cdots$$

Successive back substitutions for the u_{t-i}'s yields the following:

$$u_t = \pi_1 Z_{t-1} - \pi_2 Z_{t-2} - \cdots - \delta$$

where π_i's and δ are model parameters and are functions of Ψ_i's and μ. We can then rewrite the model as

$$Z_t = \pi_1 Z_{t-1} + \pi_2 Z_{t-2} + \cdots + \pi_p Z_{t-p} + \delta + u_t$$

which is referred to as an *autoregressive process of order p*, usually denoted as $AR(p)$. For notational convenience, we will denote the autoregressive process as shown in the following:

$$Z_t = \varphi_1 Z_{t-1} + \varphi_2 Z_{t-2} + \cdots + \varphi_p Z_{t-p} + \delta + u$$

Thus, AR processes are equivalent to MA processes of infinite order. Stationarity of AR processes is confirmed if the roots of the following characteristic equation lie outside the unit circle in the complex plane:

$$\left(1 - \varphi_1 x - \varphi_2 x^2 - \cdots - \varphi_p x^p \right) = 0$$

where x is a dummy algebraic symbol. If the process is stationary, then we should have

$$E(Z_t) = \varphi_1 E(Z_{t-1}) + \varphi_2 E(Z_{t-2}) + \cdots + \varphi_p E(Z_{t-p}) + \delta + E(u_t)$$
$$= \varphi_1 E(Z_t) + \varphi_2 E(Z_t) + \cdots + \varphi_p E(Z_t) + \delta$$
$$= E(Z_t)(\varphi_1 + \varphi_2 + \cdots + \varphi_p) + \delta$$

which yields

$$E(Z_t) = \frac{\delta}{\left(1 - \varphi_1 - \varphi_2 - \cdots - \varphi_p\right)}$$

Denoting the deviation of the process from its mean by Z_t^d, the following is obtained:

$$Z_t^d = Z_t - E(Z_t) = Z_t - \frac{\delta}{\left(1 - \varphi_1 - \varphi_2 - \cdots - \varphi_p\right)}$$
$$Z_{t-1}^d - Z_{t-1} - \frac{\delta}{\left(1 - \varphi_1 - \varphi_2 - \cdots - \varphi_p\right)}$$

Rewriting the previous expression yields

$$Z_{t-1} = Z_{t-1}^d + \frac{\delta}{\left(1 - \varphi_1 - \varphi_2 - \cdots - \varphi_p\right)}$$
$$\cdots$$
$$\cdots$$
$$Z_{t-k} = Z_{t-k}^d + \frac{\delta}{\left(1 - \varphi_1 - \varphi_2 - \cdots - \varphi_p\right)}$$

If we substitute the $AR(p)$ expression into the expression for Z_t^d, we will obtain

$$Z_t^d = \varphi_1 Z_{t-1} + \varphi_2 Z_{t-2} + \cdots + \varphi_p Z_{t-p} + \delta + u_t - \frac{\delta}{\left(1 - \varphi_1 - \varphi_2 - \cdots - \varphi_p\right)}$$

Successive back substitutions of Z_{t-j} into the preceding expression yields

$$Z_t^d = \varphi_1 Z_{t-1}^d + \varphi_2 Z_{t-2}^d + \cdots + \varphi_p Z_{t-p}^d + u_t$$

Thus, the deviation series follows the same AR process without a constant term. The tools for identifying and constructing time series models are the sample autocorrelations, r_j. For the model identification procedure, a visual assessment of the plot of r_j against j, called the sample correlogram, is used. Figure 7.2 presents examples of *sample correlograms* and the corresponding time series models.

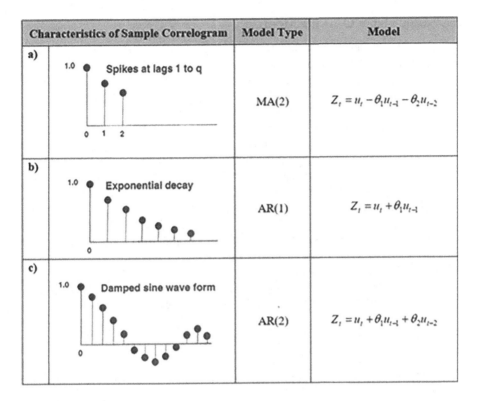

Characteristics of Sample Correlogram	Model Type	Model
a) Spikes at lags 1 to q	MA(2)	$Z_t = u_t - \theta_1 u_{t-1} - \theta_2 u_{t-2}$
b) Exponential decay	AR(1)	$Z_t = u_t + \theta_1 u_{t-1}$
c) Damped sine wave form	AR(2)	$Z_t = u_t + \theta_1 u_{t-1} + \theta_2 u_{t-2}$

FIGURE 7.2   Characteristics of sample correlogram.

A wide variety of sample correlogram patterns can be encountered in time series analysis. It is the responsibility of the analyst to choose an appropriate model to fit the prevailing time series data. Several statistical computer programs are available for performing time series analysis.

INVENTORY MANAGEMENT MODELS

Inventoried items are an important component of any supply chain, even in nonmanufacturing operations. Any resource can be viewed as an "inventoried" item for the purpose of managing the supply chain. Consequently, inventory management strategies are essential for comprehensive supply chain planning and control. Tracking activities is analogous to tracking inventory items. The important aspects of inventory management for supply systems management are:

Ability to satisfy work demands promptly by supplying materials from stock;
Availability of bulk rates for purchases and shipping;
Possibility of maintaining more stable and level resource or workforce.

Some of the basic and classical inventory control techniques are discussed in the following.

ECONOMIC ORDER QUANTITY MODEL

The economic order quantity (EOQ) model determines the optimal order quantity based on purchase cost, inventory carrying cost, demand rate, and ordering cost. The objective is to minimize the total relevant costs (TRC) of inventory. For the formulation of the model, the following notations are used:

Q is the replenishment order quantity (in units)
A is the fixed cost of placing an ordering
v is the variable cost per unit of the item to be inventoried
r is the inventory carrying charge per dollar of inventory per unit time
D is the demand rate of the item
TRC is the total relevant costs per unit time

Figure 7.3 shows the basic inventory pattern with respect to time as well as some inventory cost patterns. In the first part of the figure, one complete cycle starts from a level of Q and ends at zero inventory.

The second part of the figure shows the costs as functions of replenishment quantity. The $TRCTRC$ for order quantity Q is given by the expression as follows

$$TRC(Q) = \frac{Qvr}{2} + \frac{AD}{Q}$$

When the $TRC(Q)$ function is optimized with respect to Q, we obtain the expression for the EOQ:

$$EOQ = \sqrt{\frac{2AD}{vr}}$$

which represents the minimum TRC of inventory. The previous formulation assumes that the cost per unit is constant regardless of the order quantity. In some cases, quantity discounts may be applicable to the inventory item. The formulation for quantity discount situation is presented in the following.

QUANTITY DISCOUNT

A quantity discount may be available if the order quantity exceeds a certain level. This is referred to as the single breakpoint discount. The third part of Figure 7.3 presents the price breakpoint for quantity discount.

The unit cost is represented as shown in the following equation

$$v = \begin{cases} v_0, & 0 \leq Q < Q_b \\ v_0(1-d), & Q_b \leq Q \end{cases}$$

where
v_0 is the basic unit cost without discount
d is the discount (in decimals)

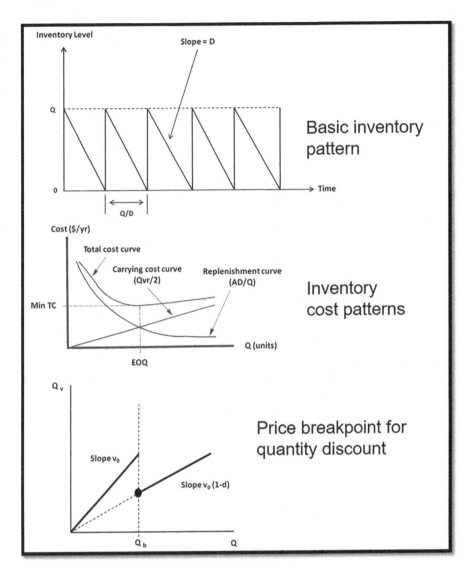

FIGURE 7.3 Inventory pattern and cost plots.

d is applied to all units when $Q \le Q_b$
Q_b is the breakpoint

CALCULATION OF TOTAL RELEVANT COST

For $0 \le Q < Q_b$, we obtain

$$TRC(Q) = \left(\frac{Q}{2}\right)v_0 r + \left(\frac{A}{Q}\right)D + Dv_0$$

For $Q_b \leq Q$, we have

$$TRC(Q)_{discount} = \left(\frac{Q}{2}\right) v_0 (1-d) r + \left(\frac{A}{Q}\right) D + D v_0 (1-d)$$

Note that for any given value of $Q, TRC(Q)_{discount} < TRC(Q)$. Therefore, if the lowest point on the $TRC(Q)_{discount}$ curve corresponds to a value of $Q^* > Q_b$ (i.e., Q is valid), then set $Q_{opt} = Q^*$.

EVALUATION OF THE DISCOUNT OPTION

The trade-off between extra carrying costs and the reduction in replenishment costs should be evaluated to see if the discount option is cost justified. Reduction in replenishment cost is composed of

1. Reduction in unit value
2. Fewer replenishments per unit time.

Case a: If reduction in acquisition costs > extra carrying costs, then set $Q_{opt} = Q_b$
Case b: If reduction in acquisition costs < extra carrying costs, then set $Q_{opt} = EOQ$ with no discount.
Case c: If Q_b is relatively small, then set $Q_{opt=EOQ}$ with discount. The optimal order quantity, Q_{opt}, can be found as follows:
  Step 1: Compute *EOQ* when d is applicable:

$$EOQ(\text{discount}) = \sqrt{\frac{2AD}{v_0 (1-d) r}}$$

Step 2: Compare $EOQ(d)$ with Q_b:

$$\text{If } EOQ(d) \geq Q_b, \text{ set } Q_{opt} = EOQ(d)$$
$$\text{If } EOQ(d) < Q_b, \text{ go to Step 3}$$

Step 3: Evaluate *TRC* for *EOQ* and Q_b:
  Suppose $d = 0.02$ and $Q_b = 100$ for the three items shown below:

Item 1: $D \ (Units/yr) = 416; v_0 (\$/Unit) = 14.20; \ (A) \ (\$) = 1.50; r(\$/\$/year) = 0.24$

Item 2: $D \ (Units/yr) = 104; v_0 (\$/Unit) = 3.10; \ (A) \ (\$) = 1.50; r(\$/\$/year) = 0.24$

Item 3: $D \ (Units/yr) = 4,160; v_0 (\$/Unit) = 2.40; \ (A) \ (\$) = 1.50; r(\$/\$/year) = 0.24$

SENSITIVITY ANALYSIS

Sensitivity analysis involves a determination of the changes in the values of a parameter that will lead to a change in a dependent variable. It is a process for determining how wrong a decision will be if assumptions on which the decision is based prove to be incorrect. For example, a "decision" may be dependent on the changes in the values of a particular parameter inventory cost may be the parameter on which the decision depends. Cost itself may depend on the values of other parameters as shown in the following:

$$\text{Sub-parameter} \rightarrow \text{Main parameter} \rightarrow \text{Decision}$$

It is of interest to determine what changes in parameter values can lead to changes in a decision. With respect to inventory management, we may be interested in the cost impact of deviation of actual order quantity from the *EOQ*. The sensitivity of cost to departures from *EOQ* is analyzed as presented in the following:

Let p represent the level of change from *EOQ*:

$$|p| \leq 1.0$$
$$Q' = (1-p)EOQ$$

Percentage cost penalty (PCP) is defined as

$$PCP = \frac{TRC(Q') - TRC(EOQ)}{TRC(EOQ)}(100)$$

$$= 50\left(\frac{p^2}{1+p}\right)$$

A plot of the sensitivity with respect to the respective percentage cost penalty (PCP) may be developed to provide a visual assessment. It can be seen that the cost is not very sensitive to minor departures from *EOQ*. We can conclude that changes within 10% of *EOQ* will not significantly affect the *TRC*. There are several inventory control algorithms available in the literature. Two examples are the Wagner–Whitin algorithm and the Silver-Meal heuristic (Badiru, 2019). The Wagner–Whitin (W–W) algorithm is an approach to deterministic inventory model. It is based on dynamic programming technique. The Silver-Meal heuristic is a simple inventory control technique that is recommended for items that have significantly variable demand pattern. Its objective is to minimize *TRC* per unit time for the duration of the replenishment quantity:

$$TRC = A + \text{Carrying costs}$$

where A is the cost of placing an order.

MODELING FOR SEASONAL PATTERN

There are numerous applications of statistical distributions to practical, real-world problems. Several distributions have been developed and successfully utilized for a large variety of problems. Despite the large number of distributions available, it is often confusing to determine which distribution is applicable to which real-world random variable. In many applications, the choices are limited to a few familiar distributions due either to a lack of better knowledge or computational ease. Such familiar distributions include the *normal, exponential,* and *uniform* distributions. To accommodate cases where the familiar distributions do not adequately represent the variable of interest, some special purpose distributions have been developed. Examples of such special distributions are *Pareto, Rayleigh, log-normal,* and *Cauchy* distributions. Consider the *cyclic probability density function* (also called *periodic distribution* or *seasonal distribution*) for special cases involving random variables that are governed by cyclic processes. These special cases include seasonal inventory control and time series processes. Instabilities in a system can be caused by erratic materials supply, cyclic inventory, and seasonal work patterns. A time series is a collection of observations that are drawn from a periodic or cyclic process. Examples of operating time series include monthly cost, quarterly revenue, and seasonal energy consumption. The standard cyclic pdf is a continuous wave-form periodic function defined on the interval [0, 2π]. The standard function is transformed into a general cyclic function defined over any time series interval [a,b]. One possible application of the cyclic distribution is the estimation of time-to-failure for components or equipment that undergoes periodic maintenance. Another application may be in the statistical analysis of the peak levels in a time series process.

The trigonometric sine function (Badiru, 2019) provides the basis for the cyclic distribution. The basic sine function is given by

$$y = \sin\theta, \quad -\infty < \theta < \infty$$

where θ is in radians. The basic sine function and its variations are shown in Figure 7.4.

The function is cyclic with period 2π. The sine function has its maxima and minima (1 and −1, respectively) at the points $\theta = \pm n\pi/2$, where n is an odd positive integer. The function also satisfies the relationship

$$\sin(\theta + 2n\pi) = \sin\theta$$

for any integer n. In general, any function $f(x)$ satisfying the relationship

$$f(x + nT) = f(x)$$

is said to be cyclic with period T, where T is a positive constant and n is an integer. A graph of $f(x)$ truncated to an interval [(a, a + T) or (a, a + T)] is called *one cycle of the function.*

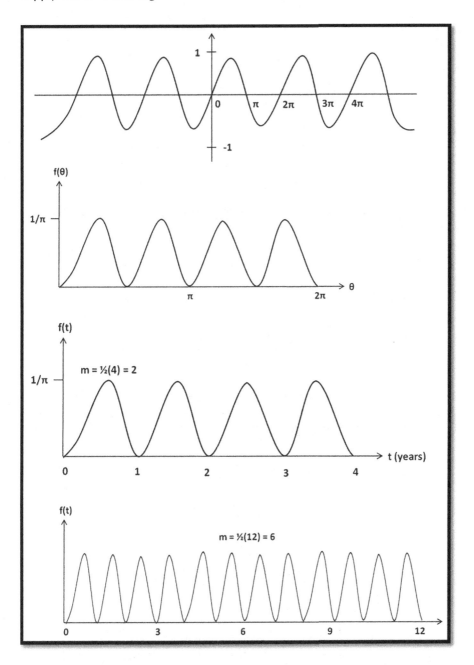

FIGURE 7.4   Basic sine function and its variations.

Standard Cyclic PDF

The cyclic pdf is defined as

$$f(x) = \begin{cases} \dfrac{1}{\pi}\sin^2 mx & 0 \le x \le 2\pi, \quad m = 1, 2, 3,\ldots \\ 0 & otherwise \end{cases}$$

where

x is in radians

m (positive integer) is the number of peaks (or modes) associated with the function on the interval $[0, \pi]$, which is half of the period

Thus, $f(x)$ represents a multimodal distribution. To be a legitimate pdf, $f(x)$ must satisfy the following conditions:

1. $(x) \ge 0$ *for allxin* $[0,\ 2\pi]$.

2. $\displaystyle\int_0^{2\pi} f(x)\,dx = 1$.

3. $P(x_1 < X < x_2) = \displaystyle\int_{x_1}^{x_2} f(x)\,dx$.

The first condition can be seen to be true by inspection. Each value of x yields a probability value equal to or greater than zero. The second condition can be verified by examining the definite integral

$$\int_0^{2\pi} f(x)\,dx = \frac{1}{\pi}\int_0^{2\pi} \sin^2(mx)\,dx$$

$$= \frac{1}{\pi}\left[\frac{x}{2} - \frac{\sin(2mx)}{4m}\right]_0^{2\pi}$$

$$= \frac{1}{\pi}\left[\pi - \frac{\sin(4mx)}{4m}\right]$$

$$= 1$$

The third condition is easily verified by computing the probability that X will fall between x_1 and x_2 as the area under the curve for $f(x)$ from x_1 to x_2.

Expected Value

The expected value of X is given by

$$E[X] = \int_0^{2\pi} xf(x)\,dx$$

$$= \frac{1}{\pi} \int_0^{2\pi} x\sin^2(mx)\,dx$$

$$= \frac{1}{\pi} \left[\frac{x^2}{4} - \frac{x\sin(2mx)}{4m} - \frac{\cos(2mx)}{8m^2} \right]_0^{2\pi}$$

$$= \pi$$

Variance

The variance of X is calculated from the theoretical definition of variance, which makes use of the definite integral from 0 to 2π.

Cumulative Distribution Function

The cumulative distribution function of $f(x)$ is given by

$$F(x) = \int_0^x f(r)\,dr$$

$$= \int_0^x \frac{1}{\pi}\sin^2(mr)\,dr$$

$$= \frac{2mx - \sin(2mx)}{4m\pi}$$

General Cyclic PDF

For practical applications, a general form of the cyclic pdf defined over any real interval $[a, b]$ will be of more interest. This general form is given by

$$f(x) = \begin{cases} \frac{1}{\pi}\sin^2 mr & 0 \le r \le 2\pi, \quad m = 1,2,3,\ldots \\ 0 & \text{otherwise} \end{cases}$$

where
    m is half of the number of peaks expected over the interval $[a, b]$
    r is the radian equivalent of the real number x in the interval $[a, b]$

The transformations from real units to radians and vice versa are accomplished by the expression as follows:

$$r = 2\pi\frac{x-a}{b-a}, \quad 0 \le r \le 2\pi, \quad a \le x \le b$$

$$x = a + (r)\frac{b-a}{2\pi}, \quad 0 \le r \le 2\pi, \quad a \le x \le b$$

Using the earlier transformation relationships and the expressions derived previously for the standard cyclic random variable, the mean and variance of the general cyclic random variable are derived to be

$$E[X] = a + \frac{b-a}{2\pi}\left(E[r]\right)$$
$$= a + \frac{b-a}{2\pi}\left(\pi\right)$$
$$= \frac{a+b}{2}$$
$$V[X] = 0 + \left(\frac{b-a}{2\pi}\right)^2 V[r]$$
$$= \frac{(b-a)^2}{4\pi^2}\left(\frac{\pi^2}{3}\right)$$
$$= \frac{(b-a)^2}{12}$$

Note that these expressions are identical to the expressions for the mean and variance of the uniform distribution. Badiru (2019) presents application examples of the sine function for seasonal inventory patterns.

Forecasting is a basic requirement for a responsive and adaptive supply chain. Coupled with inventory modeling, practical data analytics can be performed as the basis for supply chain decisions. This chapter has presented a basic analysis of forecasting and inventory modeling in the context of supply chain systems.

REFERENCES

Badiru, A. B. (2019). *Project management: Systems, principles, and applications* (2nd ed.). Taylor & Francis/CRC Press.

Badiru, A. B., Pulat, P. S., & Kang, M. (1993). DDM: Decision support system for hierarchical dynamic decision making. *Decision Support Systems*, *10*(1), 1–18.

Hogan, D., Elshaw, J., Koschnick, C., Ritschel, J., Badiru, A., & Valentine, S. (2020, October). Cost estimating using a new learning curve theory for non-constant production rates. *Forecasting*, *2*(4), 429–451.

8 Learning Curve in the Supply Chain

Reprinted with permission from:

"A Learning Curve Model Accounting for the Flattening Effect in Production Cycles" by Boone, E., Elshaw, J. J., Koschnick, C. M., Ritschel, J. D., & Badiru, A. B. (2021, January). *Defense Acquisition Research Journal, 28*(1), 72–97, Issue 95 by DAU Press.
DOI: https://doi.org/10.22594/10.22594/dau.20-850.28.01

INTRODUCTION TO THE LEARNING SYSTEM

Learning in the postal service system is like learning in any other industrial production setting. Learning and forgetting do occur in the postal system also and it can affect the framework for a flexible global supply chain. This chapter presents an analytical research of learning curves in production and supply chain systems. The outputs of production systems are the elements that go into the supply chain. Learning curves are widely used in the design of production systems because the effect of learning helps to reduce the cost per unit as production runs increase. So, there is a good connectivity between learning curves, production outputs, and the supply chain. Whether the outputs are physical products, needed services, or desired results, the same concepts of learning curves are applicable to the supply chain.

In this reprinted journal article, the authors investigate production cost estimates to identify and model modifications to a prescribed learning curve. The research created a new learning curve for production processes that incorporates a new model parameter. The new parameter allows for a steeper learning curve at the beginning of production and a flattening effect near the end of production. The new model examines the learning rate as a decreasing function over time as opposed to a constant rate that is frequently used. The purpose of this research is to determine whether a new learning curve model could be implemented to reduce the error in cost estimates for production processes. A new model was created that mathematically allows for a "flattening effect," which typically occurs later in the production process. This model was then compared to Wright's learning curve, which is a popular method used by many organizations today. The results showed a statistically significant reduction in error through the measurement of the two error terms, Sum of Squared Errors and Mean Absolute Percentage Error.

Many manufacturing firms today operate in a fiscally constrained and financially conscious environment. Managers throughout these organizations are expected to maximize the utility from every dollar as budgets and profit margins continue to shrink. Increased financial scrutiny adds greater emphasis on the accuracy of program and project management cost estimates to ensure acquisition programs are sufficiently funded. Cost estimating models and tools used by organizations must be evaluated for their relevance and accuracy to ensure reliable cost estimates. Many of the cost estimating procedures for learning curves were developed in the 1930s (Wright, 1936) and are still in use today as a primary method to model learning. As automation and robotics increasingly replace human touch-labor in the manufacturing process, the current 80-year-old learning curve model may no longer provide the most accurate approach for estimates. New learning curve methods that incorporate automated production and other factors that lead to reduced learning should be examined as an alternative for cost estimators in the acquisition process.

Since Wright's (1936) original learning curve model was developed, researchers have found other functions to model learning within the manufacturing process (Carr, 1946; Chalmers & DeCarteret, 1949; Crawford, 1944; De Jong, 1957; Towill, 1990; Towill & Cherrington, 1994). The purpose of this research is to address a gap in the literature that fails to account for the nonconstant rate of learning, which results in a flattening effect at the end of production cycles.

We investigate learning curve estimating methodology, develop learning curve theory, and pursue the development of a new estimation model that examines learning at a non-constant rate.

This research identifies and models modifications to a learning curve model such that the estimated learning rate is modeled as a decreasing learning rate function over time, as opposed to a constant learning rate that is currently in use. Wright's (1936) learning curve model in use today mathematically states that for every doubling of units there will be a constant gain in efficiency. For example, if a manufacturer observes a 10% reduction in labor hours in the time to produce unit 10 from the time to produce unit 5, then it should expect to see the same 10% reduction in labor hours in the time to produce unit 20 from the time to produce unit 10. We propose that more accurate cost estimates would result if a more flexible exponent were taken into consideration in developing the learning curve model. The proposed general modification would take the form:

$$Cost(x) = Ax^{f(x)}$$

Where:
   Cost(x) = cumulative average cost per unit
   A = theoretical cost to produce the first unit
   x = cumulative number of units produced
   f(x) = learning curve effect as a function of units produced

The exponent function above is explored in this article. Figure 8.1 demonstrates the phenomena this research examines. The black (flatter) line depicts the traditional curve where learning occurs at a constant rate; the red (steeper) line represents the hypothesized learning structure where the rate of learning changes as a function of the number of units produced; and the blue line represents notional data used to fit the two curves.

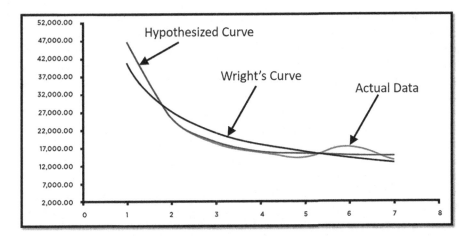

FIGURE 8.1 Learning curve depiction: Comparative curves.

To address this research gap, our study aims to model a function that has the added precision of diminishing learning effects over time by introducing a learning curve decay factor that more closely models actual production cycle learning. We accomplish this by developing a new learning curve model that minimizes the amount of error compared to current estimation models. Learning curves, specifically when estimating the expected cost per unit of complex manufactured items such as aircraft, are frequently modeled with a mathematical power function. The intent of these models is to capture the expected reduction in costs over time due to learning effects, particularly in areas with a high percentage of human touch labor. Typically, as production increases, manufacturers identify labor efficiencies and improve the process. If labor efficiencies are identified, it translates to unit cost savings over time. The general form of the learning curve model frequently used today is based on Wright's theory. Note that the structure of the exponent b ensures that as the number of units produced doubles, the cost will decrease by a given percentage defined as the learning curve slope (LCS). For example, when LCS is 0.8, then the cost per unit will decrease by 80% between units 2 and 4.

$$Cost(x) = Ax^b$$

Where:
Cost(x) = cumulative average cost per unit
A = theoretical cost to produce the first unit
x = cumulative number of units produced
$$b = \frac{\ln Learning\ Curve\ Slope}{\ln 2}$$

The cost of a particular production unit is modeled as a power function that decreases at a constant exponential rate. The problem is that the rate of decrease is not likely to be constant over time. We propose that the majority of cost improvements are to be found early in the production process, and fewer revelations are made later as the manufacturer becomes more familiar with the process. As time progresses, the production process should normalize to a steady state and additional cost reductions prove less likely. For relatively short production runs, the basic form of the learning curve may be sufficient because the hypothesized efficiencies will not have time to materialize. However, when estimating production runs over longer periods of time, the basic learning curve could underestimate the unit costs of those furthermost in the future. The underestimation occurs because the model assumes a constant learning rate, while actual learning would diminish, causing the actuals to be higher than the estimate. Current models may underestimate a significant amount when dealing with high unit cost items such as those in major acquisition programs; a small error in an estimate can be large in terms of dollars. Through the use of curve fitting techniques, a comparison can be made to determine which model best predicts learning within a particular production process. The remainder of this article is organized as follows. A literature review of the most common learning curve processes is presented in the next section, followed by methodology and model formulation. We then provide an in-depth analysis of the learning curve models, followed by future research directions, conclusions, and limitations of this research.

FROM THE LITERATURE

Learning curve research dates back to 1936, when Theodore Paul Wright published the original learning curve equation that predicted the production effects of learning. Wright recognized the mathematical relationship that exists between the time it takes for a worker to complete a single task and the number of times the worker had previously performed that task (Wright, 1936). The mathematical relationship developed from this hypothesis is that as workers complete the same process, they get better at it. Specifically, Wright realized that the rate at which they get better at that task is constant. The relationship between these two variables is as follows: As the number of units produced doubles, the worker will do it faster by a constant rate. He proposed that this relationship takes the form of:

$$F = N^x \ or \ x = \frac{Log\,F}{Log\,N},$$

"where F = a factor of cost variation proportional to the quantity N. The reciprocal of F then represents a direct percent variation of cost vs. quantity" (Wright, 1936). The relationship between these variables can be modified to predict the expected cost of a given unit number in production by multiplying the factor of cost variation by the theoretical cost of the first unit produced and is shown in Figure 8.2. It is a log linear relationship through an algebraic manipulation. The logarithmic form of this equation (taking the natural log of both sides of the equation) allows practitioners to run linear regression analysis on the data to find what slope best fits the data using a straight line (Martin, 2019).

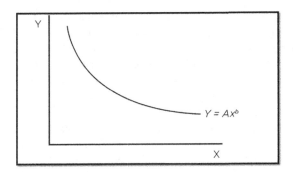

FIGURE 8.2   Wright's learning curve model.

The goal of using learning curves is to increase the accuracy of cost estimates. Having accurate cost estimates allows an organization to efficiently budget while providing as much operational capability as possible because it can allocate resources to higher priorities. While the use of learning curves focuses on creating accurate cost estimates, learning curves often use the number of labor hours it takes to perform a task. When Wright originated the theory, he proposed the output in terms of time to produce, not production cost. However, many organizations perform learning curve analysis on both production cost and time to produce, depending on the data available. Nevertheless, labor hour cost is relevant because it is based on factors such as labor rates and other associated values. The use of labor hours in learning curve development allows a common comparison over time without the effects of inflation convoluting the results. However, the same goal can be achieved by using inflation-adjusted cost values.

Wright's model has been compared to some of the more contemporary models that have surfaced in recent years since the original learning curve theory was established (Moore et al., 2015). Moore compared the Stanford-B, Dejong, and the S-Curve models to Wright's model to see if any of these functions could provide a more accurate estimate of the learning phenomenon. Both the Dejong and the S-Curve models use an incompressibility factor in the calculation. Incompressibility is a factor used to account for the percentage of automation in the production process. Values of the incompressibility factor can range from zero to one where zero is all touch labor and one is complete automation. Moore found that when using an incompressibility factor between zero and 0.1, the Dejong and S-Curve models were more accurate (Moore et al., 2015). In other words, when a production process had very little automation and high amounts of touch labor, the newer learning curve models tended to be more accurate. For all other values of incompressibility, Wright's model was more accurate.

More recently, Johnson (2016) proposed that a flattening effect is evident at the end of the production process where learning does not continue to occur at a constant rate near the end of a production cycle. Using the same models as Moore, Johnson explored the difference in accuracy between Wright's model and contemporary models early in the production process versus later in the production process. He had similar findings to Moore in that Wright's model was most accurate except in cases where the incompressibility factors were extremely low. When the incompressibility factor is low, more touch labor is involved in the process allowing for the possibility of additional learning to occur. He also found that Wright's learning curve was more accurate early in the production process whereas the Dejong and S-Curve models were more accurate later in the production process (Johnson, 2016).

Another key concept in learning curve estimation and modeling is the idea of a forgetting curve (Honious et al., 2016). A forgetting curve explains how configuration changes in the production process can cause a break in learning, which leads to loss of efficiency that had previously been gained. When a configuration change occurs, the production process changes. Changes may include factors such as using different materials, different tooling, adding steps to a process, or might even be attributed to workforce turnover. The new process affects how workers complete their tasks and causes previously learned efficiencies to be lost. If manufacturers fail to take these breaks into account, they may underestimate the total effort needed to produce a product. Honious et al. (2016) found that configuration changes significantly changed the learning curve, and that the new learning curve slope was steeper than the previous steady slope prior to a configuration change. The distinction between pre- and post-configuration change is important to ensure both types of effects are considered.

The International Cost Estimating and Analysis Association (ICEAA) published learning curve training material in 2013. While presenting the basics of learning curve theory, it also presented some rules of thumb for learning. The first rule is that learning curves are steepest when the production process is touch-labor intensive. Conversely, learning curves are the flattest when the production process is highly automated (ICEAA, 2013). Another key piece of information is that adding new work to the process can affect the overall cost. ICEAA states that this essentially adds a new curve for the added work, which increases the original curve by the amount of the new curve (ICEAA, 2013). The equation is as follows:

$$Cost(x) = A_1 x^{b_1} + A_2 \left(x - L \right)^{b_2}$$

Where:
Cost(x) = cumulative average cost per unit
A_1 = theoretical cost to produce the first unit prior to addition of new work
x = cumulative number of units produced
L = last unit produced before addition of new work
A_2 = theoretical cost to produce the first unit after addition of new work

$$b_1 = \frac{\ln \text{Learning Curve Slope prior to additional work}}{\ln 2}$$

$$b_2 = \frac{\ln \text{Learning Curve Slope after additional work}}{\ln 2} \quad \text{(typically same as } b_1)$$

The above equation is important to consider when generating an estimate after a major configuration change or engineering change proposal (ECP). For example, while producing the eighth unit of an aircraft, the customer realizes they need to drastically change the radar on the aircraft. Learning has already taken place on the first eight aircraft; the new radar has not yet been installed, and therefore no learning has taken place. To accurately consider the new learning, the radar would be treated as a second part to the equation, ensuring we account for the learning on the eight aircraft while also accounting for no learning on the new radar.

Finally, Anderlohr (1969) and Mislick and Nussbaum (2015) write about production breaks and the effects they have on a learning curve. These production breaks can cause a direct loss of learning, which can fully or partially reset the learning curve. For example, a 50% loss of learning would result in a loss of half of the cost reduction that has occurred (ICEAA, 2013). This information is important when analyzing past data to ensure that breaks in production are accounted for.

Thus far, we have laid out the fundamental building blocks for learning curve theory and how they might apply in a production environment. Wright's learning curve formula established the method by which many organizations account for learning during the procurement process. Following Wright's findings, other methods have emerged that account for breaks in production, natural loss of learning over time, incompressibility factors, and half-life analysis (Benkard, 2000). This article adds to the discussion by examining the flattening effect and how various models predict learning at different points in the production process.

When examining learning curve theory and the effects learning has on production, it is critical to understand the production process being estimated. Since Wright established learning curve theory in 1936, factory automation and technology have grown tremendously and continue to grow. Contemporary learning curve methods try to account for this automation. To get the best understanding, we examine the aircraft industry, specifically how it behaves in relation to the rest of the manufacturing industry.

The aircraft industry has relatively low automation (Kronemer & Henneberger, 1993), especially compared to other industries. Kronemer and Henneberger (1993) state that the aircraft industry is a fairly labor-intensive process with relatively little reliance on automated production techniques, despite it being a high-tech industry. Specifically, they list three main reasons why manufacturing aircraft is so labor-intensive. First, aircraft manufacturers usually build multiple models of the same aircraft, typically for the commercial sector alone. These different aircraft models mean different tooling and configurations are needed to meet the demand of the customer. Second, aircraft manufacturers deal with a very low unit volume when compared to other industries in manufacturing. The final reason for low levels of automation is the fact that aircraft are highly complex and have very tight tolerances. To attain these specifications, manufacturers must continue to use highly skilled touch laborers or spend extremely large amounts of money on machinery to replace them (Kronemer & Henneberger, 1993). For these reasons, we should typically see or use low incompressibility factors in the learning curve models when estimating within the aircraft industry.

Although the aircraft industry remains largely unaffected by the shift to machine production from human touch labor, many industries are seeing a rise in the percentage of manufacturing processes that are automated. In a *Wall Street Journal* article posted in 2012, the author showed how companies have been increasing the amount of money spent on machines and software while spending less on manpower. They proposed that part of the reason behind this shift was a temporary tax break "that allowed companies in 2011 to write off 100% of investments in the first year" (Aeppel, 2012). Tax breaks combined with extremely low interest rates provided industry with incentive to invest in future production. Investment in production technology increases the incompressibility factor that should be used when estimating the effects of learning. In a separate article for the *Wall Street Journal*, Kathleen Madigan also pointed out the increase in spending on capital investments in relation to labor. She stated that "businesses had increased their real spending on equipment and software by a strong 26%, while they have added almost nothing to their payrolls" (Madigan, 2011).

MODEL FORMULATION

Before we can begin the process of developing a new learning curve equation, we need to examine the characteristics of the curve we expected to best fit the data. Our hypothesis is that a learning curve whose slope decreases over time would fit the data better than Wright's curve. To adjust the rate at which the curve flattens, the b value from Wright's learning curve, or the exponent in the power function, needs to be adjusted. Specifically, to make the curve move in a flatter direction, the exponent in the power curve must decrease as the number of units produced increases. Initially we modified Wright's existing formula by dividing the exponent by the unit number

$$Cost(x) = Ax^{b/x}$$

Where:
    Cost(x) = cumulative average cost per unit
    A = theoretical cost of the first unit
    x = cumulative number of units produced
    b = Wright's learning curve constant

Using Wright's learning curve, b is a negative constant that has a larger magnitude for larger amounts of learning (i.e., as LCS decreases, b becomes more negative). Therefore, when b is divided by x, the amount of learning is reduced. In fact, the flattening effect is fairly drastic. For example, a standard 80% Wright's learning curve exhibits 90% learning by the second unit and flattens to 97% by the fourth unit. To implement a learning curve that has the flexibility to not flatten as quickly, we instead divide *b* by *1+x/c* where *c* is a positive constant. The term *1+x/c* is always greater than one and is increasing as *x* increases; therefore, a flattening effect always occurs (i.e., learning decreases as the number of units produced increases). The choice of the constant *c* is critical in determining how quickly the learning decreases. For example, when c = 4, a standard 80% Wright's learning curve exhibits 86% learning by the second unit and approximately 89% learning by the fourth unit. For the same standard 80% curve when c = 40, the learning decreases to 80.9% by the second unit and to 81.6% by the fourth unit. The new equation (which we also refer to as Boone's learning curve hereafter) took the form:

$$Cost(x) = Ax^{b/\left(1+\frac{x}{c}\right)}$$

Where:
Cost(x) = cumulative average cost per unit
A = theoretical cost of the first unit
x = cumulative number of units produced
b = Wright's learning curve constant
c = decay value (positive constant)

The function that modifies the traditional learning curve exponent., $1+x/c$ – has a key characteristic – ensures that the rate of learning associated with traditional learning curve theory decreases as each additional unit is produced. Specifically, $1 + x/c$ is always greater than 1 since x/c is always positive. Note that c is an estimated parameter and x increases as more units are produced, so the term x/c is decreasing. When c is large, Boone's learning curve would effectively behave like Wright's learning curve. For example, if the fitted value of c is 5,000, then $1 + x/c$ equals 1.0002 after the first unit has been produced and 1.004 after the twentieth unit has been produced. This equates to a decrease in the learning rate of the traditional theory (i.e., b) of less than 0.07%. More formally, as c goes to infinity, $b/(1 + x/c)$ goes to b.

Note that the previous discussion assumed that b was the same value for both Wright's and Boone's learning curve to help demonstrate the flattening effect. In practice, nothing precludes each of the learning curves from having different b values. For instance, if we desire a learning curve that possesses more learning early in production and less learning later in production (compared to Wright's curve), then the b parameters could be different. This is shown in Figure 8.1. In this case, Boone's curve would have a b value less than Wright's curve (i.e., a more negative value representing more learning). Then the flattening effect of dividing by $1 + x/c$ as production increases would eventually result in a curve with less learning than Wright's curve. For example, consider an 80% Wright's learning curve and a Boone's learning curve that initially has 70% learning and a decay value of 8; by the eighth production unit, Boone's curve would be at 82% learning.

EFFICACY OF SAMPLING

To test the new learning curve, we looked at quantitative data from several DoD airframes to gain a comprehensive understanding of how learning affects the cost of lot production. The costs used in this analysis are the direct lot costs and exclude costs for items such as Research, Development, Test, and Evaluation (RDT&E), support items, and spares. These data specifically include Prime Mission Equipment (PME) only as these costs are directly related to labor, and can be influenced directly through learning. To ensure we are comparing properly across time, we used inflation and rate-adjusted PME cost data for each production lot of the selected aircraft systems. The PME cost data were adjusted using escalation rates for materials using the Office of Secretary of Defense (OSD) rate tables, when applicable. We used data from fighter, bomber, and cargo aircraft, as well as missiles and munitions. This diverse dataset allowed comparison among multiple systems in different production environments.

DATA COLLECTION

Data used were gathered from the Cost Assessment Data Enterprise (CADE). CADE is a resource available to DoD cost analysts that stores historical data on weapon systems. Some of the older data also came from a DoD research library in the form of cost summary reports. The data used can be broken out by Work Breakdown Structure (WBS) or Contract Line Item Number (CLIN). For this research, the PME cost data were broken out by WBS element, then rolled up into top line, finished product elements and used for the regression analysis. In total, 46 weapon system platforms were analyzed.

ANALYSIS

Regression analysis was used to test which learning curve model was most accurate in estimating the data. The goal is to minimize the sum of squared errors (SSE) in the regression to examine how well a model estimates a given set of data. The SSE is calculated by taking the vertical distance between the actual data point (in this case lot midpoint PME cost) and the prediction line (or estimate) (Mislick & Nussbaum, 2015). This error term is then squared and the sum of these squared error terms is the value for comparing which model is a more accurate predictor. However, since an extra parameter is available in fitting the regression for the new model, it should be able to maintain or decrease the SSE in most cases. As previously mentioned, as the decay parameter approaches infinity, Boone's learning curve approaches Wright's learning curve formula. With this in mind, we also examined the Mean Absolute Percentage Error (MAPE). MAPE takes the same error term as the SSE calculation but then divides it by the actual value; then the mean of the absolute value of these modified error terms is calculated. By examining the error in terms of a percentage, comparisons between different types and sizes of systems are more robust. If Boone's curve reduces both SSE and MAPE when compared to the SSE and MAPE of Wright's curve, it would indicate the new model may be better suited for modeling learning and the associated costs.

As stated previously, Wright's learning curve is suitable for a log-log model. A log-log model is used when a logarithmic transformation of both sides of an equation results in a model that is linear in the parameters. As Wright proposed, this linear transformation occurs because learning happens at a constant rate throughout the production cycle. If learning happens at a nonconstant rate (as in Boone's learning curve), then the curve in log-log space would no longer be linear. This constraint means typical linear regression methods would not be suitable for estimating Boone's learning curve; therefore, we had to use nonlinear methods to fit these curves.

Specifically, we used the Generalized Reduced Gradient (GRG) nonlinear solver package in Excel to minimize the SSE by fitting the A, b, and c parameters. To use this solver, bounds for the three parameters had to be established. These are values that are easy to obtain for any dataset, as they are provided by Microsoft Excel when fitting a power function or by using the "linest()" function in Excel. We used this as a starting point because Wright's curve is currently used throughout the DoD. For the A variable, the lower bound was one-half of Wright's A and the upper bound was 2 times Wright's A. These values were used to give the solver model a wide enough range to avoid limiting the value but small enough to ease the search for the optimal values. Neither of these limits was found to be binding. For the exponent parameter b, we chose values between 3 and -3 times Wright's exponent value. In theory, the value of the exponent should never go above 0 due to positive learning leading to a decrease in cost, but in practice some datasets go up over time and we wanted to be able to account for those scenarios, if necessary. Again, these values between 3 and -3 times Wright's exponent value were never found to be binding limits for the model. Finally, for the decay parameter c, fitted values were bounded between 0 and $5,000$; the $5,000$ upper bound was found to be a binding constraint in the solver on several occasions. In practice, analysts could bound the value as high as possible to reduce error, but in the case of this research, we used $5,000$ as no significant change was evidenced from $5,000$ to infinity – relaxing this bound would have only further reduced the SSE for Boone's learning curve.

STATISTICAL SIGNIFICANCE TESTING

Once the SSE and MAPE values were calculated for each learning curve equation, we tested for significance to determine whether the difference between the error values for the two equations were statistically different. Specifically, we conducted t-tests on the differences in error terms between Wright's and Boone's learning curve equations. This t-test was conducted for both SSE and MAPE values separately. A nonsignificant t-test indicates no statistically significant difference between the two learning curves.

ANALYSIS AND RESULTS

The table given in an article by Boone et al., published in 2021, shows the SSE and MAPE values for both Wright's and Boone's learning curve for each system in the dataset. The last two columns are the percentage difference in SSE and MAPE between the two learning curve methods. This percentage was calculated by taking the difference of Boone's error term minus Wright's error term divided by Wright's error term. Negative values represent programs where Boone's learning curve had less error than Wright's learning curve, and positive values represent programs where Wright's curve had less error than Boone's curve.

Based on this analysis, we observed that Boone's learning curve reduced the SSE in approximately 84% of programs and reduced MAPE in 67% of programs. The mean reduction of SSE and MAPE was 27% and 17%, respectively. As previously mentioned, these values were based on using both learning curve equations to minimize the SSE for each system in the dataset. This is standard practice in the DoD as prescribed by the U.S. Government Accountability Office (GAO, 2009) *Cost Estimating and Assessment Guide* when predicting the cost of subsequent units or subsequent lots.

We conducted additional tests to determine if a statistical difference existed between the means of both curve estimation techniques. On average, programs estimated using Boone's learning curve had a lower error rate (M = 4.73, SD = 2.15) than those estimated using Wright's learning curve (M = 8.64, SD = 4.55). Additionally, the difference between these two error rates expressed as a percentage and compared to a hypothesized value of 0 (no difference) was significant, $t(46) = -4.87$, p < .0001, and represented an effect of d = 1.10. We then applied the same test to the difference in the MAPE values from Boone's learning curve and Wright's learning curve. On average, programs estimated using Boone's learning curve had a lower MAPE value (M = .08, SD = .07) than those estimated using Wright's MAPE value (M = .10, SD = .06). The difference between these two estimates has a mean of -.17, which translates to Boone's curve reducing MAPE by 17% more on average. Additionally, the difference between these two error rates expressed as a percentage and compared to a hypothesized value of 0 (no difference) was significant, $t(46) = -3.48$, p < .0005, and represented an effect of d = .22. The results indicate that in both SSE and MAPE, Boone's learning curve reduced the error, and that each of those values was statistically significant when using an alpha value of 0.05.

DISCUSSION

As stated previously, an average of a 27% reduction in the SSE resulted from among the 46 programs analyzed. These results were statistically significant. Also, a 17% reduction in the MAPE resulted from among the programs analyzed, which was also found to be statistically significant. Based on these results, we can conclude that Boone's learning curve equation was able to reduce the overall error in cost estimates for our sample. This information is critical to allow the DoD to calculate more accurate cost estimates and better allocate its resources. These conclusions help answer our three guiding research questions. Specifically, we were looking for the point where Wright's model became less accurate than other models. We found that adding a decay factor caused the learning curve to flatten out over time, which resulted in less error than Wright's model. Additionally, we found that Boone's learning curve was more accurate throughout the entire production process, not just during the tail end when production was winding down. Boone's learning curve was steeper during the early stages of production when it's hypothesized that the most learning occurs. Toward the end of the production process, Boone's curve flattens out more than Wright's curve, supporting our contention that learning toward the end of the production cycle yields diminishing returns. While Wright's curve assumes constant learning throughout the entire process, Boone's curve treats learning in a nonlinear fashion that slows down over time. By reducing the error in the estimates and properly allocating resources, the DoD could potentially minimize risk for all parties involved. The benefit of Boone's learning curve is accuracy in the estimation process. If labor estimates aren't accurate in the production process, risks escalate, such as schedule slip, cost overruns, and increased costs for all involved. Accuracy in the cost estimate should be the goal of both the contractor and government, thereby facilitating the acquisition process with better data.

LIMITATIONS

One limitation of this study is that all 46 of the weapon systems analyzed were U.S. Air Force systems. While the list included many platforms spanning decades, we hesitate to draw conclusions outside of the U.S. Air Force without further research and analysis. That said, we see no reason our model wouldn't apply equally well in any aircraft production environment, both within and outside the DoD. Another limitation in this research is the use of PME cost as opposed to labor hours. Labor-hour data are not readily available across many platforms, which led to the use of PME cost. Contractor data provided to the government normally come in the form of lots, which is the lowest level tracked by cost estimators. To compare learning curves across multiple platforms, the same level of analysis is required to ensure a fair comparison. Future research should attempt to examine data at the individual level of analysis between systems and exclude those where only lot data are available. Because there are inherently less lots than units, this may affect how the equation behaves when applied at the unit level. For this research, we used the lot midpoint formula/method (Mislick & Nussbaum, 2015), but further research should be conducted to evaluate the performance of Boone's learning curve with unitary data. Finally, we only performed a comparison to Wright's learning curve since that is a primary method of estimation in the DoD. A comparison with other learning curve models may yield different results, although previous research found those curves were not statistically better than Wright's.

RECOMMENDATIONS FOR FUTURE RESEARCH

Data outside of the U.S. Air Force should be examined to test whether this equation applies broadly to programs, and not just to Air Force programs. Also, conducting the analysis with unitary data could confirm that this works for predicting subsequent units as well as subsequent lots, while reducing error over Wright's method. We also tried to select weapon systems that had minimal automation in the production process. However, DeJong's Learning Formula is another derivation from Wright's original in which an incompressibility factor is introduced. The incompressibility factor represents the amount of automation in the production process, which limits how much learning can occur (Badiru et al., 2013). Other models such as the S-Curve model (Carr, 1946) and a more recent version (Towill, 1990; Towill & Cherrington, 1994) also account for some form of incompressibility. Additional research could also include modifications to Boone's formula to try and further reduce the error types listed in this research. Furthermore, fitting Boone's curve in this analysis was based on past data whereas cost estimates are used to project future costs. Therefore, future research should identify decay values for different types of weapon systems – similar to the way learning curve rates are established for different categories of programs. Finally, further research could examine whether the incorporation of multiple learning curve equations at different points in the production process would be beneficial to reducing additional error in the estimates.

We developed a new learning curve equation utilizing the concept of learning decay. This equation was tested against Wright's learning equation to see which equation provided the least amount of error when looking at both the SSE and MAPE. We found that Boone's learning curve reduced error in both cases and that this reduction in error was statistically significant. Follow-on research in this field could lead to further discoveries and allow for broader use of this equation in the cost community.

LEARNING CURVE IMPLICATION FOR SUPPLY CHAINS

Although the methodology of this research centers on general production systems using Air Force data, it is directly applicable to the case of general relationships between a production system and a supply chain networks. The half-life learning curve model presented by Badiru (2012) is particularly relevant for dynamic technology-sensitive supply chains, where learning and re-learning can change over a short period of time. An appreciation and incorporation of learning curves can help mitigate supply chain risks.

REFERENCES

Aeppel, T. (2012, January 17). Man vs. machine: Behind the jobless recovery. *Wall Street Journal*. www.wsj.com/articles/SB10001424052970204468004577164710231081398

Anderlohr, G. (1969). What production breaks cost. *Industrial Engineering*, *1*(9), 34–36.

Badiru, A. B. (2012). Half-life learning curves in the defense acquisition life cycle. *Defense Acquisition Research Journal*, *19*(3), 283–308. www.afit.edu/BIOS/publications/HalflifeLearningCurvesinDefenseAcquisitionLifeCycleBadiruDARJ2012.pdf

Badiru, A. B., Elshaw, J., & Mack, E. (2013). Half-life learning curve computations for airframe life-cycle costing of composite manufacturing. *Journal of Aviation and Aerospace Perspective*, *3*(2), 6–37.

Benkard, C. L. (2000). Learning and forgetting: The dynamics of aircraft production. *American Economic Review*, *90*(4), 1034–1054. www.aeaweb.org/articles?id=10.1257/aer.90.4.1034

Carr, G. W. (1946). Peacetime cost estimating requires new learning curves. *Aviation*, *45*(4), 220–228.

Chalmers, G., & DeCarteret, N. (1949). *Relationship for determining the optimum expansibility of the elements of a peacetime aircraft procurement program*. USAF Air Materiel Command.

Crawford, J. R. (1944). *Estimating, budgeting and scheduling*. Lockheed Aircraft Co.

De Jong, J. R. (1957). The effects of increasing skill on cycle time and its consequences for time standards. *Ergonomics*, *1*(1), 51–60. https://doi.org/10.1080/00140135708964571

Honious, C., Johnson, B., Elshaw, J., & Badiru, A. (2016, May 4–5). *The impact of learning curve model selection and criteria for cost estimation accuracy in the DoD: Maj Gen C. Blake, USAF (Chair)*. Proceedings of the 13th Annual Acquisition Research Symposium, Monterey, CA. https://apps.dtic.mil/dtic/tr/fulltext/u2/1016823.pdf

International Cost Estimating and Analysis Association. (2013). *Cost estimating body of knowledge* [PowerPoint slides]. ICEAA. www.iceaaonline.com/cebok/

Johnson, B. J. (2016). *A comparative study of learning curve models and factors in defense cost estimating based on program integration, assembly, and checkout* [Master's thesis, Air Force Institute of Technology]. https://apps.dtic.mil/dtic/tr/fulltext/u2/1056447.pdf

Kronemer, A., & Henneberger, J. E. (1993). Productivity in aircraft manufacturing. *Monthly Labor Review, 16*(6), 24–33. www.bls.gov/mfp/mprkh93.pdf

Madigan, K. (2011, September 28). It's man vs. machine and man is losing. *Wall Street Journal.* https://blogs.wsj.com/economics/2011/09/28/its-man-vs-machine-and-man-is-losing/

Martin, J. R. (2019). *What is a learning curve?* Management and Accounting Web. Retrieved June 15, 2019, from http://maaw.info/LearningCurveSummary.htm

Mislick, G. K., & Nussbaum, D. A. (2015). *Cost estimation: Methods and tools.* Wiley. www.academia.edu/24430098/Cost_Estimation_Methods_and_Tools_by_Gregory_K_Mislick_and_Daniel_A_Nussbaum_1st_Edition

Moore, J. R., Elshaw, J., Badiru, A. B., & Ritschel, J. D. (2015, October). Acquisition challenge: The importance of incompressibility in comparing learning curve models. *Defense Acquisition Research Journal, 22*(4), 416–449. www.dau.edu/library/arj/ARJ/ARJ75/ARJ75-ONLINE-FULL.pdf

Towill, D. R. (1990). Forecasting learning curves. *International Journal of Forecasting, 6*(1), 25–38. www.sciencedirect.com/science/article/abs/pii/016920709090095S

Towill, D. R., & Cherrington, J. E. (1994). Learning curve models for predicting the performance of AMT. *The International Journal of Advanced Manufacturing Technology, 9,* 195–203. https://doi.org/10.1007/BF01754598

U.S. Government Accountability Office. (2009). *GAO cost estimating and assessment guide (GAO-09-3SP).* Government Printing Office. www.gao.gov/new.items/d093sp.pdf

Wright, T. P. (1936). Factors affecting the cost of airplanes. *Journal of the Aeronautical Sciences, 3*(4), 122–128. https://arc.aiaa.org/doi/10.2514/8.155

9 Supply Chain Supplier

The performance of a supply chain is ultimately a function of the inputs going into the supply chain. These inputs originate from the supplier sources that produce physical products, desired services, or needed services. If the source can be optimized, the process can be improved, and the supply output can be enhanced. When supply disruptions occur, as we saw in the case of the COVID-19 pandemic, the origin of problems can be from any point in the overall spectrum of the supply chain. It is in this regard that this chapter is dedicated to providing the quantitative methodology of optimizing supplier selection. This chapter is based on Ravindran and Wadhwa (2009). Full details of the approach can be found in Wadhwa and Ravindran (2007), Ravindran and Wadhwa (2009), Ravindran (2008), Ravindran and Warsing (2017), and all the references therein.

SUPPLIER SELECTION PROBLEM

A supply chain consists of a connected set of activities concerned with planning, coordinating, and controlling materials, parts, and finished goods from supplier to customer (Ravindran & Wadhwa, 2009). The contribution of the purchasing function to the profitability of the supply chain has assumed greater proportions in recent years; one of the most critical functions of purchasing is selection of suppliers. For most manufacturing firms, the purchasing of raw material and component parts from suppliers constitutes a major expense. Raw material cost accounts for 40%–60% of production costs for most US manufacturers. In fact, for the automotive industry, the cost of components and parts from outside suppliers may exceed 50% (of sales (Ravindran & Wadhwa, 2009). For technology firms, purchased materials and services account for 80% of the total production cost. It is vital to the competitiveness of most firms to be able to keep the purchasing cost to a minimum. In today's competitive operating environment it is impossible to successfully produce low-cost, high-quality products without good suppliers. A study carried out by the Aberdeen Group (Ravindran & Wadhwa, 2009) found that more than 83% of the organizations engaged in outsourcing achieved significant reduction in purchasing cost, more than 73% achieved reduction in transaction cost, and over 60% were able to shrink sourcing cycles.

Supplier selection process is difficult because the criteria for selecting suppliers could be conflicting. Figure 9.1 illustrates the various factors which could impact the supplier selection process (Ravindran & Wadhwa, 2009). Supplier selection is a multiple criteria optimization problem that requires trade-off among different qualitative and quantitative factors to find the best set of suppliers. For example, the supplier with the lowest unit price may also have the lowest quality. The problem is

DOI: 10.1201/9781032620701-9

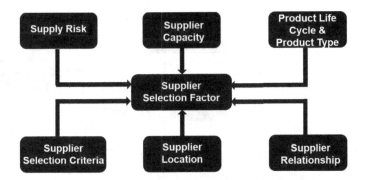

FIGURE 9.1 Supplier selection factors.

Source: Adapted from Ravindran & Wadhwa (2009).

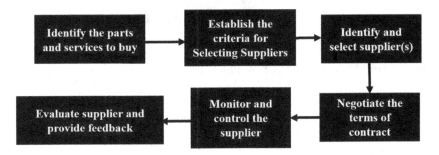

FIGURE 9.2 Supplier selection steps.

Source: Adapted from Ravindran & Wadhwa (2009).

also complicated by the fact that several conflicting criteria must be considered in the decision-making process.

Most of the time, buyers have to choose among a set of suppliers by using some predetermined criteria, such as quality, reliability, technical capability, lead-times, and so forth, even before building long-term relationships. To accomplish these goals, two basic and interrelated decisions must be made by a firm. The firm must decide which suppliers to do business with and how much to order from each supplier. Weber et al. (Ravindran & Wadhwa, 2009) refer to this pair of decisions as the supplier selection problem.

Figure 9.2 illustrates the steps in the supplier selection process. The first step is to determine whether to *make or buy* the item. Most organizations buy those parts which are not core to the business or not cost-effective if produced in-house. The next step is to define the various criteria for selecting the suppliers. The criteria for selecting a supplier of critical product may not be the same as a supplier of maintenance, repairs, and operations (MRO) items. Once a decision to buy the item is taken, the most critical step is selecting the right supplier. Once the suppliers are chosen the organization has to negotiate terms of contract and monitor their performance.

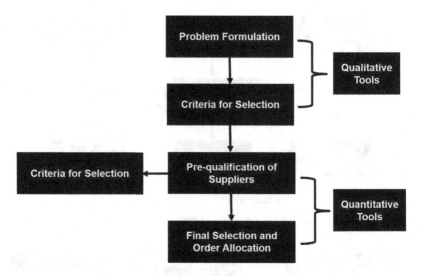

FIGURE 9.3   Supplier selection process.

Source: Adapted from Ravindran & Wadhwa (2009).

Finally, the suppliers have to be constantly evaluated, and the feedback should be provided to purchasing and the suppliers.

SUPPLIER SELECTION MODELS

As mentioned earlier, the supplier selection activity plays a key role in cost reduction and is one of the most important functions of the purchasing department. Different mathematical, statistical, and game theoretical models have been proposed to solve the problem. References in Ravindran and Wadhwa (2009) provide an overview of supplier selection methods.

De Boer et al. (Ravindran & Wadhwa, 2009) stated that that supplier selection process is made up of several decision-making steps as shown in Figure 9.3.

PROBLEM FORMULATION

Decision support methods for problem formulation are those methods that support the purchasing manager in identifying the need for supplier selection; in this case problem formulation involves determining what the ultimate problem is and why supplier selection is the better choice. According to a survey by Aissaoui et al. (Ravindran & Wadhwa, 2009) and De Boer et al. (Ravindran & Wadhwa, 2009), there are no papers that deal with the problem formulation.

CRITERIA FOR SELECTION

Depending on the buying situation, different sets of criteria may have to be employed. Criteria for supplier selection have been studied extensively since the

1960s. Dickson's study (Ravindran & Wadhwa, 2009) was the earliest to review the supplier selection criteria. He identified 23 selection criteria with varying degrees of importance for the supplier selection process. Dickson's study was based on a survey of purchasing agents and managers. A follow-up study was done by Weber et al. (Ravindran & Wadhwa, 2009). They reviewed 74 articles published in the '70s and '80s and categorized them based on Dickson's 23 selection criteria. In their studies, net price, quality, delivery, and service were identified as the key selection criteria. They also found that under just-in-time manufacturing, quality and delivery became more important than net price. Supplier selection criteria may also change over time. Wilson (Ravindran & Wadhwa, 2009) examined studies conducted in 1974, 1982, and 1993 on the relative importance of the selection criteria. She found that quality and service criteria began to dominate price and delivery. She concluded that globalization of the marketplace and the increasingly competitive atmosphere had contributed to the shift.

De Boer et al. (Ravindran & Wadhwa, 2009) pointed out that there are only two known techniques for formulating supplier selection criteria. Mandal and Deshmukh (Ravindran & Wadhwa, 2009) proposed interpretive structural modeling (ISM) as technique based on group judgment to identify and summarize relationships between supplier choice criteria through a graphical model. Vokurka et al. (Ravindran & Wadhwa, 2009) developed an expert system for the formulation of the supplier selection criteria. Nominal group technique involving all the stakeholders of the supplier selection decision can also be used to identify the important supplier selection criteria (Ravindran & Wadhwa, 2009).

PRE-QUALIFICATION OF SUPPLIERS

Pre-qualification is the process of screening suppliers to identify a suitable subset of suppliers. Pre-qualification reduces the large set of initial suppliers to a smaller set of acceptable suppliers and is more of a sorting process. De Boer et al. (Ravindran & Wadhwa, 2009) have cited many different techniques for prequalification. Some of these techniques are categorical methods, data envelopment analysis (DEA), cluster analysis, case-based reasoning (CBR) systems, and multi-criteria decision-making methods (MCDM). Several authors have worked on prequalification of suppliers. Weber and Ellram (Ravindran & Wadhwa, 2009) and Weber et al. (Ravindran & Wadhwa, 2009) have developed DEA methods for pre-qualification. Hinkel et al. (Ravindran & Wadhwa, 2009) and Holt (Ravindran & Wadhwa, 2009) used cluster analysis for pre-qualification. Ng and Skitmore (Ravindran & Wadhwa, 2009) developed CBR systems for pre-qualification. Finally, Mendoza et al. (Ravindran & Wadhwa, 2009) developed a three-phase multi-criteria method to solve a general supplier selection problem. The paper combines analytic hierarchy process (AHP) with goal programming for both pre-qualification and final order allocation.

FINAL SELECTION

Most of the publications in the area of supplier selection have focused on final selection. In the final selection step, the buyer identifies the suppliers to do business with and allocates order quantities among the chosen supplier(s). In reviewing the literature, there are basically two kinds of supplier selection problem, as stated by Ghodsypour and O'Brien (Ravindran & Wadhwa, 2009):

- *Single sourcing*, which implies that any one of the suppliers can satisfy the buyer's requirements of demand, quality, delivery, and so on;
- *Multiple sourcing*, which implies that there are some limitations in suppliers' capacity, quality, and so on, and multiple suppliers have to be used.

Specifically, no one supplier can satisfy the buyer's total requirements and the buyer needs to purchase some part of its demand from one supplier, and the other part from another supplier to compensate for the shortage of capacity or low quality of the first supplier. Several methods have been proposed in the literature for single sourcing as well as for multiple sourcing.

Single sourcing models: Single sourcing is a possibility when a relatively small number of parts are procured (Ravindran & Wadhwa, 2009). Some of the methods used in single sourcing are:

- *Linear weighted point*: The linear weighted point method is the most widely used approach for single sourcing. This approach uses a simple scoring method, which heavily depends on human judgment. Some of the references that discuss this approach include Wind and Robinson (Ravindran & Wadhwa, 2009) and Zenz (Ravindran & Wadhwa, 2009).
- *Cost ratio*: Cost ratio is a more complicated method in which the cost of each criterion is calculated as a percentage of the total purchased value, and a net adjusted cost for each supplier is determined. This approach is complex and needs a lot of financial information. This approach was proposed by Timmerman (Ravindran & Wadhwa, 2009).
- *Analytic hierarchy process* (AHP): AHP, developed by Saaty (Ravindran & Wadhwa, 2009) in the early '80s, is a good tool to solve multiple criterion problems with finite alternatives. It is a powerful and flexible decision-making process to help the decision maker set priorities and make the best decision when both qualitative and quantitative aspects of decision are to be considered. AHP has been extensively used for supplier selection problems (Ravindran & Wadhwa, 2009).
- *Total cost of ownership* (TCO): TCO is a method which looks beyond the price of a product to include other related costs like quality costs, technology costs, and so on. Finally, the business is awarded to the supplier with lowest unit total cost. General Electric Wiring Devices have developed a total cost supplier selection method that takes into account risk factors, business desirable factors, and measurable cost factors (Ravindran & Wadhwa, 2009). The TCO approach has also been used by Ellram (Ravindran & Wadhwa, 2009) and Degraeve (Ravindran & Wadhwa, 2009).

MULTIPLE SOURCING MODELS

Multiple sourcing can offset the risk of supply disruption. In multiple sourcing, a buyer purchases the same item from more than one supplier. Mathematical programming is the most appropriate method for multiple sourcing decisions. It allows the buyers to consider different constraints while choosing the suppliers and their order allocation. Two types of mathematical programming models are found in the literature: single objective and multi-objective models.

Single Objective Models

Moore and Fearon (Ravindran & Wadhwa, 2009) stated that price, quality, and delivery are important criteria for supplier selection. They discussed the use of linear programming in decision making.

Gaballa (Ravindran & Wadhwa, 2009) applied mathematical programming to supplier selection in a real case. He used a mixed integer programming to formulate a decision-making model for the Australian Post Office. The objective for this approach is to minimize the total discounted price of allocated items to the suppliers. Anthony and Buffa (Ravindran & Wadhwa, 2009) developed a single objective linear programming model to support strategic purchasing scheduling (SPS). The linear model minimized the total cost by considering limitations of purchasing budget, supplier capacities, and buyer's demand. Price and storage cost were included in the objective function. The costs of ordering, transportation, and inspection were not included in the model. Bender et al. (Ravindran & Wadhwa, 2009) applied single objective programming to develop a commercial computerized model for supplier selection at IBM. They used mixed integer programming to minimize the sum of purchasing, transportation, and inventory costs. Narasimhan and Stoynoff (Ravindran & Wadhwa, 2009) applied a single objective, mixed integer programming model to a large manufacturing firm in the Midwest to optimize the allocation procurement for a group of suppliers. Turner (Ravindran & Wadhwa, 2009) presented a single objective linear programming model for British Coal. This model minimized the total discounted price by considering the supplier capacity, maximum and minimum order quantities, demand, and regional allocated bounds as constraints. Pan (Ravindran & Wadhwa, 2009) proposed multiple sourcing for improving the reliability of supply for critical materials, in which more than one supplier is used and the demand is split between them. The author used a single objective linear programming model to choose the best suppliers, in which three criteria are considered: Price, quality, and service. Seshadri et al. (Ravindran & Wadhwa, 2009) developed a probabilistic model to represent the connection between multiple sourcing and its consequences, such as number of bids, the seller's profit, and the buyer's price. Benton (Ravindran & Wadhwa, 2009) developed a nonlinear programming model and a heuristic procedure using Lagrangian relaxation for supplier selection under conditions of multiple items, multiple suppliers, resource limitations, and quantity discounts. Chaudhry et al. (Ravindran & Wadhwa, 2009) developed linear and mixed integer programming models for supplier selection. Price, delivery, quality, and quantity discount were included in their model. Papers by Degraeve (Ravindran & Wadhwa, 2009) and Ghodsypour and O'Brien (Ravindran & Wadhwa, 2009) tackled the supplier selection issue in the framework of total cost of ownership or total cost of logistics. Jayaraman et al. (Ravindran & Wadhwa, 2009) formulated a mixed integer linear programming model for solving the supplier selection problem with multiple products, buyers, and suppliers. Feng et al. (Ravindran & Wadhwa, 2009) presented a stochastic integer-programming model for simultaneous selection of tolerances and suppliers based on the quality loss function and process capability index.

MULTI-CRITERIA MODELS

Among the different multi-criteria approaches for supplier selection, goal programming (GP) is the most commonly used method. Buffa and Jackson (Ravindran & Wadhwa, 2009) presented a multi-criteria linear goal programming model. In this model two sets of factors are considered: Supplier attributes, which include quality, price, service experience, early, late, and on-time deliveries; and the buying firm's specifications, including material requirement and safety stock. Sharma et al. (Ravindran & Wadhwa, 2009) proposed a GP formulation for attaining goals pertaining to price, quality, and lead-time under demand and budget constraints. Liu et al. (Ravindran & Wadhwa, 2009) developed a decision support system by integrating AHP with linear programming. Weber et al. (Ravindran & Wadhwa, 2009) used multi-objective linear programming for supplier selection to systematically analyze the trade-off between conflicting factors. In this model aggregate price, quality, and late delivery were considered as goals. Karpak et al. (Ravindran & Wadhwa, 2009) used visual interactive goal programming for the supplier selection process. The objective was to identify and allocate order quantities among suppliers while minimizing product acquisition cost and maximizing quality and reliability. Bhutta and Huq (Ravindran & Wadhwa, 2009) illustrated and compared the technique of total cost of ownership and AHP in supplier selection process. Wadhwa and Ravindran (Ravindran & Wadhwa, 2009) formulated a supplier selection problem with price, lead-time, and quality as three conflicting objectives. The suppliers offered quantity discounts and the model was solved using goal programming, compromise programming, and weighted objective methods.

MULTI-CRITERIA RANKING METHODS FOR SUPPLIER SELECTION

Many organizations have a large pool of suppliers to select from. The supplier selection problem can be solved in two phases. The first phase reduces the large number of candidate suppliers to a manageable size. In the second phase, a multiple criteria optimization model is used to allocate order quantities among the short-listed suppliers.

PRE-QUALIFICATION OF SUPPLIERS

Pre-qualification is defined as the process of reducing the large set of suppliers to a smaller manageable number by ranking the suppliers under a predefined set of criteria. The primary benefits of pre-qualification of suppliers are as follows (Ravindran & Wadhwa, 2009):

1. The possibility of rejecting good suppliers at an early stage is reduced.
2. Resource commitment of the buyer towards purchasing process is optimized.
3. With the application of pre-selected criteria, the pre-qualification process is rationalized.

In this section, we present multiple criteria ranking approaches for the supplier selection problem, namely, the pre-qualification of suppliers.

In the pre-qualification process (Phase 1), readily available qualitative and quantitative data are collected for the various suppliers. This data can be obtained from trade journals, the Internet, and past transactions, to name a few. Once this data is gathered, these suppliers are evaluated using multiple criteria ranking methods. The decision maker then selects a portion of the suppliers for extensive evaluation in Phase 2.

The first step in pre-qualification is defining the selection criteria. We have used the following 14 pre-qualification criteria as an illustration. The prequalification criteria have been split into various subcategories such as organizational criteria, experience criteria, and so on. The various prequalification criteria are described below:

Organizational Criteria:

Size of company (C1): Size of the company can be either its number of employees or its market capitalization;
Age of company (C2): Age of the company is the number of years that the company has been in business;
R&D activities (C3): Investment in research and development.

Experience Criteria:

Project type (C4): Specific types of projects completed in the past;
Project size (C5): Specific sizes of projects completed in the past.

Delivery Criteria:

Accuracy (C6): Meeting the promised delivery time;
Capacity (C7): Capacity of the supplier to fulfill orders;
Lead-time (C8): Supplier's promised delivery lead-time.

Quality Criteria:

Responsiveness (C9): If there is an issue concerning quality, how fast the supplier reacts to correct the problem;
Defective rate (C10): Rate of defective items among orders shipped.

Cost Criteria:

Order change and cancellation charges (C11): Fee associated with modifying or changing orders after they have been placed;
Unit cost: Price per item (C12).

Miscellaneous Criteria:

Labor relations (C13): Number of strikes or any other labor problems encountered in the past;
Procedural compliances (C14): Conformance to national/international standards (e.g. ISO 9000).

For the sake of illustration, we assume there are 20 suppliers during pre-qualification. The 14 supplier criteria values for the initial set of 20 suppliers are presented by Ravindran & Wadhwa (2009). For all the supplier criteria, larger values are preferred. Next, we discuss several multiple criteria ranking methods for short listing the suppliers. Each method has advantages and limitations. The methods that we discuss here are:

1. L_p metric method
2. Rating method
3. Borda count
4. Analytic hierarchy process (AHP)
5. Cluster analysis.

For a more detailed discussion of multi-criteria ranking methods, the reader is referred to Ravindran (Ravindran & Wadhwa, 2009).

USE OF L_p METRIC FOR RANKING SUPPLIERS

Mathematically, the L_p metric represents the distance between two vectors **x** and **y**, where $x,y \in R^n$, and is given by:

$$\|x - y\|_p = \left[\sum_{j=1}^{n} | x_j - y_j |^p \right]^{1/p}$$

One of the most commonly used L_p metrics is the L_2 metric (p = 2), which measures the Euclidean distance between two vectors. The ranking of suppliers is done by calculating the L_p metric between the ideal solution (H) and each vector representing the supplier's ratings for the criteria. The ideal solution represents the best values possible for each criterion from the initial list of suppliers. Since no supplier will have the best values for all criteria (e.g. a supplier with minimum cost may have poor quality and delivery time), the ideal solution is an artificial target and cannot be achieved. The L_p metric approach computes the distance of each supplier's attributes from the ideal solution and ranks the supplier's based on that distance (the smaller the better).

RATING (SCORING) METHOD

Rating is one of the simplest and most widely used ranking methods under conflicting criteria. First, an appropriate rating scale is agreed to (e.g. from 1 to 10, where 10 is the most important and 1 is the least important selection criteria). The scale should be clearly understood to be used properly. Next, using the selected scale, the decision maker (DM) provides a rating r_j for each criterion, C_j. The same rating can be given to more than one criterion. The ratings are then normalized to determine the weights of the criteria j. Assuming n criteria:

$$W_j = \frac{r_j}{\sum_{j=1}^{j=n} r_j} \quad for \quad j = 1, 2, \ldots \ldots n$$

$$Note \sum_{j=1}^{n} w_j = 1$$

Next, a weighted score of the attributes is calculated for each supplier as follows:

$$S_k = \sum_{j=1}^{n} W_j f_{jk} \quad for \; k = 1 \ldots \ldots K \; \text{where} \; f_{jk}\text{'s are the criteria values for supplier k.}$$

The suppliers are then ranked based on their scores. The supplier with the highest score is ranked first. The rating method requires relatively little cognitive burden on the DM.

BORDA COUNT

This method is named after Jean Charles de Borda, an 18th-century French physicist. The method is as follows:

- The n criteria are ranked 1 (most important) to n (least important).
- Criterion ranked 1 gets n points, 2nd rank gets (n − 1) points, and the last place criterion gets 1 point.
- Weights for the criteria are calculated as follows:
  - Criterion ranked 1 = n/S
  - Criterion ranked 2 = (n − 1)/S
  - Last criterion = 1/S

where S is the sum of all the points = n(n + 1)/2.

ANALYTIC HIERARCHY PROCESS

The analytic hierarchy process (AHP), developed by Saaty (Ravindran & Wadhwa, 2009), is a multi-criteria decision-making method for ranking alternatives. Using AHP, the decision maker can assess not only quantitative but also various intangible factors such as financial stability, feeling of trust, and so on in the supplier selection process. The buyer establishes a set of evaluation criteria and AHP uses these criteria to rank the different suppliers. AHP can enable DM to represent the interaction of multiple factors in complex and unstructured situations.

BASIC PRINCIPLES OF AHP

- Design a hierarchy: Top vertex is the main objective and bottom vertices are the alternatives. Intermediate vertices are criteria/subcriteria (which are more and more aggregated as you go up in the hierarchy).

- At each level of the hierarchy, a paired comparison of the vertices criteria/subcriteria is performed from the point of view of their "contribution (weights)" to each of the higher-level vertices to which they are linked.
- Uses both rating method and comparison method. A numerical scale 1–9 (1 = equal importance; 9 = most important).
- Uses pairwise comparison of alternatives with respect to each criterion (subcriterion) and gets a numerical score for each alternative on every criterion (subcriterion).
- Compute total weighted score for each alternative and rank the alternatives accordingly.

STEPS OF THE AHP MODEL

Step 1: In the first step, carry out a pairwise comparison of criteria using the 1–9 degree of importance scale (see Ravindran & Wadhwa, 2009).

If there are n criteria to evaluate, then the pairwise comparison matrix for the criteria is given by; $A_{(NxN)} = [a_{ij}]$, where a_{ij} represents the relative importance of criterion i with respect to criterion j. Set $a_{ii} = 1$ and $a_{ji} = \dfrac{1}{a_{ij}}$.

Step 2: Compute the normalized weights for the main criteria. We obtain the weights using L_1 norm. The two step process for calculating the weights is as follows:

- Normalize each column of $A_{(NxN)}$ using L_1 norm.

$$r_{ij} = \frac{a_{ij}}{\sum\limits_{i=1}^{n} a_{ij}}$$

- Average the normalized values across each row.

$$w_i = \frac{\sum\limits_{j=1}^{n} r_{ij}}{n}$$

Step 3: In the third step we check for consistency of the pairwise comparison matrix using the eigenvalue theory as follows (Ravindran & Wadhwa, 2009).

- Using the pairwise comparison matrix A and the weights W compute AW. Let the vector $X = (X_1, X_2, X_3 \ldots X_n)$ denote the values of AW.
- Compute

$$\lambda_{max} = Average \left[\frac{X_1}{W_1}, \frac{X_2}{W_2}, \frac{X_3}{W_3} \ldots \ldots \frac{X_n}{W_n} \right]$$

- Consistency Index (CI) is given by

$$CI = \frac{\lambda_{max} - n}{n - 1}$$

Saaty (Ravindran & Wadhwa, 2009) generated a number of random positive reciprocal matrices with $a_{ij} \in (1, 9)$ for different sizes and computed their average CI values, denoted by RI, as given below. He defines the consistency ratio (CR) as $CR = \dfrac{CI}{RI}$. If $CR < 0.15$, then accept the pairwise comparison matrix as consistent.

Step 4: In the next step, we compute the relative importance of the sub-criteria in the same way as done for the main criteria. Step 2 and Step 3 are carried out for every pair of sub-criteria with respect to their main criterion. The final weights of the sub-criteria are the product of the weights along the corresponding branch.

Step 5: Repeat Steps 1, 2 and 3 and obtain:

- Pairwise comparison of alternatives with respect to each criterion using the ratio scale (1–9).
- Normalized scores of all alternatives with respect to each criterion. Here, an (mxm) matrix S is obtained, where S_{ij} = normalized score for alternative "*i*" with respect to criterion '*j*' and *m* is the number of alternatives.

Step 6: Compute the total score (TS) for each alternative as follows $TS_{(mx1)} = S_{(mxn)} W_{(nx1)}$, where W is the weight vector obtained after Steps 3 and 4. Using the total scores, the alternatives are ranked. There is commercially available software for AHP called Expert Choice.

CLUSTER ANALYSIS

Cluster analysis (CA) is a statistical technique particularly suited to grouping of data. It is gaining wide acceptance in many different fields of research such as data mining, marketing, operations research, and bioinformatics. CA is used when it is believed that the sample units come from an unknown population. Clustering is the classification of similar objects into different groups, or more precisely, the partitioning of a data set into subsets (clusters), so that the data in each subset share some common trait. CA develops sub-sets of the raw data such that each sub-set contains member of like nature (similar supplier characteristics) and that difference between different sub-sets is as pronounced as possible.

There are two types of clustering algorithms (Ravindran & Wadhwa, 2009):

- *Hierarchical*: Algorithms which employ hierarchical clustering find successive clusters using previously established clusters. Hierarchical algorithms can be further classified as *agglomerative* or *divisive*. Agglomerative algorithms begin with each member as a separate cluster and merge them into successively larger clusters. On the other hand, divisive algorithms begin with the whole set as one cluster and proceed to divide it into successively smaller clusters. The agglomerative method is the most common hierarchical method.
- *Partitional*: In partitional clustering, the algorithm finds all the clusters at once. An example of partitional methods is k-means clustering.

PROCEDURE FOR CLUSTER ANALYSIS

Clustering process begins with formulating the problem and concludes with carrying out analysis to verify the accuracy and appropriateness of the method. The clustering process has the following steps:

1. Formulate the problem and identify the selection criteria.
2. Decide on the number of clusters.
3. Select a clustering procedure.
4. Plot the dendrogram (a tree diagram used to illustrate the output of clustering analysis) and carry out analysis to compare the means across various clusters.

Let us illustrate the cluster analysis using our supplier selection example.

Step 1: In the first step every supplier is rated on a scale of 0–1 for each attribute.

Step 2: In this step we need to decide on the number of clusters. We want the initial list of suppliers to be split into two categories, good suppliers and bad suppliers; hence the number of clusters is two.

Step 3: Next we apply both hierarchical and partitional clustering methods to supplier data. We choose the method which has the highest R-sq value pooled over all the 14 attributes.

The R-sq value for k-means is the highest among different methods; hence k-means is chosen for clustering. There are several other methods available for determining the goodness of fit (Ravindran & Wadhwa, 2009).

COMPARISON OF RANKING METHODS

Different ranking methods can provide different solutions resulting in rank reversals. In extensive empirical studies with human subjects, it has been found (Ravindran & Wadhwa, 2009) that Borda count (with pairwise comparison of criteria) rankings are generally in line with AHP rankings. Given the increased cognitive burden and expensive calculations required for AHP, Borda count might be selected as an appropriate method for supplier rankings. Even though the rating method is easy to use, it could lead to several ties in the final rankings, thereby making the results less useful.

GROUP DECISION MAKING

Most purchasing decisions, including the ranking and selection of suppliers, involve the participation of multiple DMs and the ultimate decision is based on the aggregation of DM's individual judgments to arrive at a group decision. The rating method, Borda count, and AHP discussed in this section can be extended to group decision making as described below:

1. *Rating method*: Ratings of each DM for every criterion is averaged. The average ratings are then normalized to obtain the group criteria weights.
2. *Borda count*: Points are assigned based on the number of DMs that assign a particular rank for a criterion. These points are then totaled for each criterion

and normalized to get criteria weights. (This is similar to how the college polls are done to get the top 25 football or basketball teams.)

3. *AHP*: There are two methods to get the group rankings using AHP.

   a. *Method 1*: Strength of preference scores assigned by individual DMs are aggregated using geometric means and then used in the AHP calculations.

   b. *Method 2*: First, all the alternatives are ranked by each DM using AHP. The individual rankings are then aggregated to a group ranking using Borda count.

MULTI-OBJECTIVE SUPPLIER ALLOCATION MODEL

As a result of pre-qualification, the large number of initial suppliers is reduced to a manageable size. In the second phase of the supplier selection, detailed quantitative data such as price, capacity, quality, and so on are collected on the shortlisted suppliers and are used in a multi-objective framework for the actual order allocation. We consider multiple buyers, multiple products, and multiple suppliers with volume discounts. The scenario of multiple buyers is possible in the case of a central purchasing department, where different divisions of an organization buy through one purchasing department. Here the number of buyers will be equal to the number of divisions buying through central purchasing. In all other cases, the number of buyers is equal to one. We consider the least restrictive case where any of the buyers can acquire one or more products from any suppliers, namely, a multiple sourcing model.

In this phase of the supplier selection process, an organization will make the following decisions:

- To choose the most favorable suppliers who would meet its supplier selection criteria for the various components;
- To order optimal quantities from the chosen most favorable suppliers to meet its production plan or demand.

The mathematical model for the order allocation problem is discussed next.

NOTATIONS USED IN THE MODEL

Model Indices

I Set of products to be purchased
J Set of buyers who procure multiple units in order to fulfill some demand
K Potential set of suppliers
M Set of incremental price breaks for volume discounts

Model Parameters

p_{ikm} Cost of acquiring one unit of product i from supplier k at price level m
b_{ikm} Quantity at which incremental price breaks occurs for product i by supplier k

$\mathbf{F_k}$ Fixed ordering cost associated with supplier k

$\mathbf{d_{ij}}$ Demand of product i for buyer j

$\mathbf{l_{ijk}}$ Lead time of supplier k to produce and supply product i to buyer j. The lead time of different buyers could be different because of geographical distances.

$\mathbf{q_{ik}}$ Quality that supplier k maintains for product i, which is measured as percent of defects

$\mathbf{CAP_{ik}}$ Production capacity for supplier k for product i

$\mathbf{N}$ Maximum number of suppliers that can be selected

Decision Variables in the Model

$\mathbf{X_{ijkm}}$ Number of units of product i supplied by supplier k to buyer j at price level m

$\mathbf{Z_k}$ Denotes if a particular supplier is chosen or not. This is a binary variable which takes a value 1 if a supplier is chosen to supply any product and is 0 if the supplier is not chosen at all.

$\mathbf{Y_{ijkm}}$ This is a binary variable which takes a value 1 if price level m is used and 0 otherwise

MATHEMATICAL FORMULATION OF THE ORDER ALLOCATION PROBLEM

The conflicting objectives used in the model are simultaneous minimization of price, lead-time, and rejects. It is relatively easy to include other objectives also. The mathematical form for these objectives is:

1. Price (z_1): Total cost of purchasing has two components: Fixed and variable cost.

   Total variable cost: The total variable cost is the cost of buying every additional unit from the suppliers and is given by:

   $$\sum_i \sum_j \sum_k \sum_m P_{ikm} \cdot X_{ijkm}$$

   Fixed cost: If a supplier k is used then there is a fixed cost associated with it, which is given by:

   $$\sum_k F_k \cdot Z_k.$$

   Hence the total purchasing cost is:

   $$\sum_i \sum_j \sum_k \sum_m P_{ikm} \cdot X_{ijkm} + \sum_k F_k \cdot Z_k$$

2. Lead-time (z_2): $\displaystyle\sum_i \sum_j \sum_k \sum_m l_{ijk} \cdot X_{ijkm}$

   The product of lead-time of each product and quantity supplied is summed over all the products, buyers and suppliers and should be minimized.

3. Quality (z_3): $\sum_i \sum_j \sum_k \sum_m q_{ik} \cdot X_{ijkm}$

The product of rejects and quantity supplied is summed over all the products, buyers and suppliers and should be minimized. Quality in our case is measured in terms of percentage of rejects.

The constraints in the model are as follows:

1. *Capacity constraint*: Each supplier k has a maximum capacity for product i, CAP_{ik}. Total order placed with this supplier must be less than or equal to the maximum capacity. Hence the capacity constraint is given by:

$$\sum_i \sum_j \sum_m X_{ijkm} \leq (CAP_{ik})Z_k \quad \forall k$$

The binary variable on the right hand side of the constraint implies that a supplier cannot supply any products if not chosen (i.e., if Z_k is 0).

2. *Demand constraint*: The demand of buyer j for product i has to be satisfied using a combination of the suppliers. The demand constraint is given by:

$$\sum_k \sum_m X_{ijkm} = d_{ij} \quad \forall i, j$$

3. *Maximum number of suppliers*: The maximum number of suppliers chosen must be less than or equal to the specified number. Hence this constraint takes the following form:

$$\sum_k Z_k \leq N$$

4. *Linearizing constraints*: In the presence of incremental price discounts, objective function is non-linear. The following set of constraints are used to linearize it:

$$X_{ijkm} \leq (b_{ikm} - b_{ikm-1}) * Y_{ijkm} \quad \forall i, j, k, 1 \leq m \leq m_k$$
$$X_{ijkm} \geq (b_{ikm} - b_{ikm-1}) * Y_{ijkm+1} \quad \forall i, j, k, 1 \leq m \leq m_k - 1$$

$0 = b_{i,k,0} < b_{i,k,1} < \dots \dots \dots < b_{i,k,m_k}$ is the sequence of quantities at which price break occurs. p_{ikm} is the unit price of ordering X_{ijkm} units from supplier k at level m, if $b_{i,k,m-1} < X_{ijkm} \leq b_{i,k,m} (1 \leq m \leq m_k)$.

The above constraints force quantities in the discount range for a supplier to be incremental. Because the "quantity" is incremental, if the order quantity lies in discount interval m, namely, $Y_{ijkm} = 1$, then the quantities in interval 1 to m − 1, should be at the maximum of those ranges. Constraint (16) also assures that a quantity in any range is no greater than the width of the range.

5. *Non-negativity and binary constraint*: $X_{ijkm} \geq 0; Z_k, Y_{ijkm} \in (0,1)$

GOAL PROGRAMMING METHODOLOGY

One way to treat multiple criteria is to select one criterion as primary and the other criteria as secondary. The primary criterion is then used as the optimization objective function, while the secondary criteria are assigned acceptable minimum and maximum values and are treated as problem constraints. However, if careful considerations were not given while selecting the acceptable levels, a feasible solution that satisfies all the constraints may not exist. This problem is overcome by *goal programming*, which has become a popular practical approach for solving multiple criteria optimization problems.

Goal programming (GP) falls under the class of methods that use completely pre-specified preferences of the decision maker in solving the multi-criteria mathematical programming problems. In goal programming, all the objectives are assigned target levels for achievement and a relative priority on achieving those levels. Goal programming treats these targets as *goals to aspire for* and not as absolute constraints. It then attempts to find an optimal solution that comes as "close as possible" to the targets in the order of specified priorities. In this section, we shall discuss how to formulate goal programming models and their solution methods.

Before we discuss the formulation of goal programming problems, we discuss the difference between the terms *real constraints* and *goal constraints* (or simply *goals*) as used in goal programming models. The real constraints are absolute restrictions on the decision variables, while the goals are conditions one would like to achieve but are not mandatory. For instance, a real constraint given by

$$x_1 + x_2 = 3$$

requires all possible values of $x_1 + x_2$ to always equal 3. As opposed to this, a goal requiring $x_1 + x_2 = 3$ is not mandatory, and we can choose values of $x_1 + x_2 \geq 3$ as well as $x_1 + x_2 \leq 3$. In a goal constraint, positive and negative deviational variables are introduced to represent constraint violations as follows:

$$x_1 + x_2 + d_1^- - d_1^+ = 3 \qquad d_1^+, d_1^- \geq 0$$

Note that, if $d_1^- > 0$, then $x_1 + x_2 < 3$, and if $d_1^+ > 0$, then $x_1 + x_2 > 3$.

By assigning suitable weights w_1^- and w_1^+ on d_1^- and d_1^+ in the objective function, the model will try to achieve the sum $x_1 + x_2$ as close as possible to 3. If the goal were to satisfy $x_1 + x_2 \geq 3$, then only d_1^- is assigned a positive weight in the objective, while the weight on d_1^+ is set to zero.

GENERAL GOAL PROGRAMMING MODEL

A general multiple criteria mathematical programming (MCMP) problem is given as follows:

$$\text{Max} \quad \mathbf{F}(\mathbf{x}) = \{f_1(\mathbf{x}), f_2(\mathbf{x}), ..., f_k(\mathbf{x})\}$$

$$\text{Subject to } g_j(\mathbf{x}) \leq 0 \text{ for } j = 1, ..., m$$

where $\mathbf{x}$ is an n-vector of *decision variables* and $f_i(\mathbf{x})$, $i = 1, ..., k$ are the k *criteria/objective functions.*

$$\text{Let } S = \{\mathbf{x} \, / \, g_j(\mathbf{x}) \leq 0, \text{ for all 'j'}\}$$
$$Y = \{\mathbf{y} \, / \, \mathbf{F}(\mathbf{x}) = \mathbf{y} \text{ for some } \mathbf{x} \in S\}$$

S is called the *decision space* and Y is called the *criteria or objective space* in MCMP.

Consider the general MCMP problem presented earlier. The assumption that there exists an optimal solution to the MCMP problem involving multiple criteria implies the existence of some preference ordering of the criteria by the decision maker (DM). The goal programming (GP) formulation of the MCMP problem requires the DM to specify an acceptable level of achievement (b_i) for each criterion f_i and specify a weight w_i (ordinal or cardinal) to be associated with the deviation between f_i and b_i. Thus, the GP model of an MCMP problem becomes:

$$\text{Minimize } Z = \sum_{i=1}^{k} (w_i^+ \, d_i^+ + w_i^- \, d_i^-)$$

$$\text{Subject to: } f_i(x) + d_i^- - d_i^+ = b_i \text{ for } i = 1, ..., k$$
$$g_j(x) \leq 0 \text{ for } j = 1, ..., m$$
$$x_j, d_i^-, d_i^+ \geq 0 \text{ for all } i \text{ and } j$$

The above equation represents the objective function of the GP model, which minimizes the weighted sum of the deviational variables. The system of represents the goal constraints relating the multiple criteria to the goals/targets for those criteria. The variables, d_i^- and d_i^+ are the deviational variables, representing the underachievement and overachievement of the ith goal. The set of weights (w_i^+ and w_i^-) may take two forms:

1. Pre-specified weights (cardinal)
2. Preemptive priorities (ordinal).

Under pre-specified (cardinal) weights, specific values in a relative scale are assigned to w_i^+ and w_i^- representing the DM's "trade-off" among the goals. Once w_i^+ and w_i^- are specified, the goal program reduces to a single objective optimization problem. The cardinal weights could be obtained from the DM using any of the methods discussed earlier (rating, Borda count, and AHP). However, for this method to work effectively, criteria values have to be scaled properly. In reality, goals are usually incompatible (i.e., in commensurable), and some goals can be achieved only at the expense of some other goals. Hence, preemptive goal programming, which is more common in practice, uses ordinal ranking or preemptive priorities to the goals by assigning incommensurable goals to different priority levels and weights to goals at the same priority level. In this case, the objective function of the GP model takes the form

$$\text{Minimize } Z = \sum_p P_p \sum_i (w_{ip}^+ d_i^+ + w_{ip}^- d_i^-)$$

where P_p represents priority p with the assumption that P_p is much larger then P_{p+1} and w_{ip}^+ and w_{ip}^- are the weights assigned to the ith deviational variables at priority p. In this manner, lower priority goals are considered only after attaining the higher priority goals. Thus, preemptive goal programming is essentially a sequential single objective optimization process, in which successive optimizations are carried out on the alternate optimal solutions of the previously optimized goals at higher priority. In addition to preemptive and non-preemptive goal programming models, other approaches (fuzzy GP, min-max GP) have also been proposed.

PREEMPTIVE GOAL PROGRAMMING

For the three criteria supplier order allocation problem, the preemptive GP formulation will be as follows:

$$\min P_1 d_1^+ + P_2 d_2^+ + P_3 d_3^+ \tag{27}$$

Subject to.

$$\sum_i \sum_j \sum_k \sum_m l_{ijk} \cdot x_{ijkm} + d_1^- - d_1^+ = \text{Lead time goal} \tag{28}$$

$$\left(\sum_i \sum_j \sum_k \sum_m P_{ikm} \cdot x_{ijkm} + \sum_k F_k \cdot z_k \right) + d_2^- - d_2^+ = \text{Price goal} \tag{29}$$

$$\sum_i \sum_j \sum_k \sum_m q_{ik} \cdot x_{ijkm} + d_3^- - d_3^+ = \text{Quality goal} \tag{30}$$

$$d_n^-, d_n^+ \geq 0 \quad \forall n \in \{1,...,3\} \tag{31}$$

$$\sum_j \sum_m x_{ijkm} \leq CAP_{ik} \cdot z_k \quad \forall i,k \tag{32}$$

$$\sum_k \sum_m x_{ijkm} = D_{ij} \quad \forall i,j \tag{33}$$

$$\sum_k z_k \leq N \tag{34}$$

$$x_{ijkm} \leq (b_{ikm} - b_{ik(m-1)}) \cdot y_{ijkm} \quad \forall i,j,k \quad 1 \leq m \leq m_k \tag{35}$$

$$x_{ijkm} \geq (b_{ikm} - b_{ik(m-1)}) \cdot y_{ijk(m+1)} \quad \forall i,j,k \quad 1 \leq m \leq m_k - 1 \tag{36}$$

$$x_{ijkm} \geq 0 \quad z_k \in \{0,1\} \quad y_{ijkm} \in \{0,1\} \tag{37}$$

NON-PREEMPTIVE GOAL PROGRAMMING

In the non-preemptive GP model, the buyer sets goals to achieve for each objective and preferences in achieving those goals expressed as numerical weights. In the non-preemptive GP model, the buyer has three goals, namely,

- Limit the lead-time to Lead goal with weight w_1.
- Limit the total purchasing cost to Price goal with weight w_2.
- Limit the quality to Quality goal with weight w_3.

The weights w_1, w_2 and w_3 can be obtained using the methods discussed. The non-preemptive GP model can be formulated as

$$\text{Min } Z = w_1 * d_1^+ + w_2 * d_2^+ + w_3 * d_3^+$$

Subject to the constraints to the earlier constraints.

In the above model d_1^+, d_2^+ and d_3^+ represent the overachievement of the stated goals. Due to the use of the weights, the model needs to be scaled. The weights w_1, w_2 and w_3 can be varied to obtain different goal programming optimal solutions.

TCHEBYCHEFF (MIN–MAX) GOAL PROGRAMMING

In this GP model, the DM only specifies the goals/targets for each objective. The model minimizes the maximum deviation from the stated goals. For the supplier selection problem the Tchebycheff goal program becomes:

$\text{Min Max } \left(d_1^+, d_2^+, d_3^+ \right)$

$$d_i^+ \geq 0 \forall i$$

The above can be reformulated as a linear objective by setting

$$\text{Max } (d_1^+, d_2^+, d_3^+) = M \geq 0$$

Thus, we have the following equivalent equation:

$$\text{Min } Z = M$$

Subject to:

$$M \geq \left(d_1^+ \right)$$

$$M \geq \left(d_2^+ \right)$$

$$M \geq \left(d_3^+ \right)$$

$$d_i^+ \geq 0 \forall i$$

The constraints stated earlier will also be included in this model. The advantage of the Tchebycheff goal program is that there is no need to get preference information (priorities or weights) about goal achievements from the DM. Moreover, the problem reduces to a single objective optimization problem. The disadvantages of this method are (i) the scaling of goals is necessary (as required in non-preemptive GP) and (ii) outliers are given more importance and could produce poor solutions.

FUZZY GOAL PROGRAMMING

Fuzzy goal programming uses the ideal values as targets and minimizes the maximum normalized distance from the ideal solution for each objective. An ideal solution is the vector of best values of each criterion obtained by optimizing each criterion independently ignoring other criteria. In this example, ideal solution is obtained by minimizing price, lead-time, and quality independently. In most situations the ideal solution is an infeasible solution since the criteria conflict with one another.

If M equals the maximum deviation from the ideal solution, then the fuzzy goal programming model is as follows:

$$\text{Min } Z = M$$

Subject to:

$$M \geq \left(d_1^+\right) / \lambda_1$$
$$M \geq \left(d_2^+\right) / \lambda_2$$
$$M \geq \left(d_3^+\right) / \lambda_3$$
$$d_i^+ \geq 0 \ \forall i$$

The constraints stated earlier will also be included in this model. In the above model λ_1, λ_2, and λ_3 are scaling constants to be set by the user. A common practice is to set the values λ_1, λ_2, and λ_3 equal to the respective ideal values. The advantage of fuzzy GP is that no target values have to be specified by the DM.

For additional readings on the variants of fuzzy GP models, the reader is referred to Ignizio and Cavalier (Ravindran & Wadhwa, 2009), Tiwari et al. (Ravindran & Wadhwa, 2009), Mohammed (Ravindran & Wadhwa, 2009), and Hu et al. (Ravindran & Wadhwa, 2009).

An excellent source of reference for goal programming methods and applications is the textbook by Schniederjans (Ravindran & Wadhwa, 2009).

See Ravindran & Wadhwa (2009) for a case study that illustrates the four goal programming methods using a supplier order allocation.

While supplier selection plays an important role in purchasing, it is especially important for cases of supply disruption, such as the COVID-19 pandemic. This chapter illustrates the use of both discrete and continuous multi-criteria decision-making techniques to optimize the supplier selection process. In this chapter, we present the supplier selection problem in two phases. In the first phase, called pre-qualification, we reduce the initial set of large number suppliers to a manageable set. Phase one reduces the effort of the buyer and makes the pre-qualification process entirely objective. In the second phase, we analyze the shortlisted suppliers using the multi-objective technique known as goal programming. We consider several conflicting criteria, including, price, lead-time and quality. An important distinction of multi-objective techniques is that it does not provide one optimal solution but a number of solutions known as efficient solutions. Hence, the role of the decision maker (buyer) is more important than before. By involving the decision maker early in the process, the acceptance of the model results by management becomes easier. The efficient solutions are compared using the value path approach to show the criteria trade-off obtained using different goal programming approaches. Besides goal programming, there are other approaches to solve the multi-criteria optimization model for the supplier selection problem. They include the weighted objective method, compromise programming, and interactive approaches. Interested readers can refer to Wadhwa and Ravindran (Ravindran & Wadhwa, 2009) and Ravindran (Ravindran & Wadhwa, 2009) for more details. Ravindran & Wadhwa (2009) also provides information on the computer software available for MCDM methods.

The supplier selection models can be extended in many different directions. One of the areas is managing supplier risk. Along with cost, quality, technology, and service criteria used in sourcing decisions, there is a need to integrate global risk factors such as political stability, currency fluctuations, taxes, local content rules, infrastructure (ports, raw materials, communication, transportation, certification, pandemic lockdown, etc.) in the supplier selection process. Another area of research is *supplier monitoring*. Since the supplier performance factors can change over time, real-time monitoring of suppliers becomes critical. The monitoring issues are to determine what supplier performance factors to monitor and the frequency of monitoring.

REFERENCES

Ravindran, A. R. (2008). *Operations research and management science handbook*. CRC Press.

Ravindran, A. R., & Wadhwa, V. (2009). Multiple criteria optimization models for supplier selection. In A. B. Badiru & M. Thomas (Eds.), *Handbook of military industrial engineering*. CRC Press/Taylor and Francis.

Ravindran, A. R., & Warsing, D. P., Jr. (2017). *Supply chain engineering: Models and applications*. CRC Press/Taylor and Francis.

Wadhwa, V., & Ravindran, A. R. (2007). Vendor selection in outsourcing. *Computers & Operations Research, 34*, 3725.

10 Global Supply Chain Under Climate Change

IRREFUTABILITY OF CLIMATE CHANGE

Whether we accept it or not, climate change is impacting everything we do in commerce, industry, business, government, and private enterprises. Extreme weather phenomena ranging from heat domes to dust storms are affecting the global supply chain. As a climate-combat response, the need for a global flexible supply chain is heightened. Saudi Arabia reports that over 1,000 pilgrims died of heat-related impacts and related bodily stress during the 2024 Hajj pilgrimage in Mecca. Temperatures hit the highest level in two decades. In as much as our operations involve the movement of items on the global platform, our supply chains will be subject to weather disruptions. Only the postal system has been globally successful and effective in the movement of goods and services across the various obstacles created by climate change. More rains, more winds, more fires, more dust storms, and more environmental turbulence in every region of our planet create impediments for the global supply chain. Industrial engineering modeling and optimization is required to respond and adapt to the climate changes as they impact the supply chain. One case example is the fresh fish supply chain in the town of Epe, Lagos State, Nigeria. From firsthand knowledge, the market has been decimated incrementally by the adversities posed by climate over the past recent decades. Figure 10.1 illustrates the environmental degradation, while Figure 10.2 shows some of the struggles faced at the fish market level.

COMMUNICATION, COOPERATION, AND COORDINATION FOR CLIMATE RESPONSE

Ultimately, getting things done, in the context of environmental response and sustainability, requires the involvement of many participants. Such participation requires communication, cooperation, and coordination. This chapter introduces the Triple C principle of project execution based on stages of communication, cooperation, and coordination (Badiru, 2008). As presented by the Chinese proverb below, involvement of every team member is critical for overall success of a project.

> Tell me, and I forget;
> Show me, and I remember;
> Involve me, and I understand.
> —Chinese proverb

The Triple C model facilitates better understanding and involvement based on foundational communication. The Triple C approach elucidates the integrated involvement of communication, cooperation, and coordination. Communication is the foundation

DOI: 10.1201/9781032620701-10

FIGURE 10.1   Climate-impacted fishing scenario at Epe, Lagos State, Nigeria.

FIGURE 10.2   Epe fish market, Lagos State, Nigeria.

for cooperation, which in turn is the foundation for coordination. Communication leads to cooperation, which leads to coordination, which leads to project harmony, which leads to project success.

The primary lesson of the Triple C model is not to take cooperation for granted. It must be pursued, solicited, and secured explicitly. The process of securing cooperation requires structured communication upfront. It is only after cooperation is in effect that all project efforts can be coordinated.

The Triple C model has been used effectively in practice to enhance project performance because most project problems can be traced to initial communication problems. The Triple C approach works because it is very simple – simple to understand and simple to implement. The simplicity comes from the fact that most of the required elements of the approach are already being done within every organization, albeit in a non-structured manner. The Triple C model puts the existing processes into a structural approach to communication, cooperation, and coordination.

The idea for the Triple C model originated from a complex facility redesign project (Badiru et al., 1993) conducted for Tinker Air Force Base (TAFB) in Oklahoma City by the School of Industrial Engineering, University of Oklahoma, from 1985 through 1989. The project was a part of a reconstruction project following a disastrous fire that occurred in the base's repair/production facility in November 1984. The urgency, complexity, scope ambiguity, confusion, and disjointed directions that existed in the early days of the reconstruction effort led to the need to develop a structured approach to communication, cooperation, and coordination of the various work elements. In spite of the high pressure timing of the project, the author called a Time-out-of-Time (TOOT) so that a process could be developed for project communication leading to personnel cooperation, and eventually facilitating task coordination. The investment of TOOT time resulted in a remarkable resurgence of cooperation where none existed at the beginning of the project. Encouraged by the intrinsic occurrence of cooperation, the process was further enhanced and formalized as the Triple C approach to the project's success. The approach was credited with the overall success of the project. The qualitative approach of Triple C complemented the technical approaches used on the project to facilitate harmonious execution of tasks. Many projects fail when the stakeholders get too wrapped up into the technical requirements at the expense of qualitative requirements. Other elements of "C," such as collaboration, commitment, and correlation, are embedded in the Triple C structure. Of course, the constraints of time, cost, and performance must be overcome all along the way.

Organizations thrive by investing in three primary resources: *people* who do the work, the *tools* that the people use to do the work, and the *process* that governs the work that the people do. Of the three, investing in people is the easiest thing an organization can do, and we should do it whenever we have an opportunity. The Triple C approach incorporates the qualitative (human) aspects of a project into overall project requirements.

The Triple C model was first used in 1985 and subsequently introduced in print in 1987 (Badiru, 1987). The project scenario that led to the development of the Triple Model was later documented in Badiru et al. (1993). The model is an effective project planning and control tool. The model states that project management

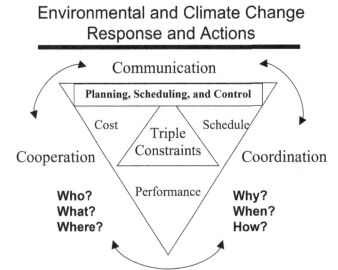

FIGURE 10.3   Triple C for communication, cooperation, and coordination.

can be enhanced by implementing it within the integrated functions summarized below:

- Communication
- Cooperation
- Coordination.

The model facilitates a systematic approach to project planning, organizing, scheduling, and control. The Triple C model is distinguished from the 3C approach commonly used in military operations. The military approach emphasizes personnel management in the hierarchy of command, control, and communication. This places communication as the last function. The Triple C, by contrast, suggests communication as the first and foremost function. The Triple C model can be implemented for project planning, scheduling, and control purposes.

Figure 10.3 shows the application of Triple C for project planning, scheduling, and control within the confines of the triple constraints of cost, schedule, and performance. Each of these three primary functions of project management requires effective communication, sustainable cooperation, and adaptive coordination.

Triple C illustrates the basic questions of what, who, why, how, where, and when. It highlights what must be done and when. It can also help to identify the resources (personnel, equipment, facilities, etc.) required for each effort. It points out important questions such as the following:

- Does each project participant know what the objective is?
- Does each participant know their role in achieving the objective?
- What obstacles may prevent a participant from playing their role effectively?

Triple C can mitigate disparity between idea and practice because it explicitly solicits information about the critical aspects of a project in terms of the following queries:

TYPES OF COMMUNICATION

- Verbal;
- Written;
- Body language;
- Visual tools (e.g., graphical tools);
- Sensual (use of all five senses: Sight, smell [olfactory], touch [tactile], taste, hearing [auditory]);
- Simplex (unidirectional);
- Half-duplex (bi-directional with time lag);
- Full-duplex (real-time dialogue);
- One-on-one;
- One-to-many;
- Many-to-one.

TYPES OF COOPERATION

- Proximity
- Functional
- Professional
- Social
- Romantic
- Power influence
- Authority influence
- Hierarchical
- Lateral
- Cooperation by intimidation
- Cooperation by enticement.

TYPES OF COORDINATION

- Teaming
- Delegation
- Supervision
- Partnership
- Token-passing
- Baton hand-off.

TRIPLE C QUESTIONS

Questioning is the best approach to getting information for effective project management. Everything should be questioned. By upfront questions, we can preempt and avert project problems later on. Typical questions to ask under the Triple C approach are:

- What is the purpose of the project?
- Who is in charge of the project?
- Why is the project needed?
- Where is the project located?
- When will the project be carried out?
- How will the project contribute to increased opportunities for the organization?
- What is the project designed to achieve?
- How will the project affect different groups of people within the organization?
- What will be the project approach or methodology?
- What other groups or organizations will be involved (if any)?
- What will happen at the end of the project?
- How will the project be tracked, monitored, evaluated, and reported?
- What resources are required?
- What are the associated costs of the required resources?
- How do the project objectives fit the goal of the organization?
- What respective contribution is expected from each participant?
- What level of cooperation is expected from each group?
- Where is the coordinating point for the project?

The key to getting everyone on board with a project is to ensure that task objectives are clear and comply with the principle of **SMART** as outlined below:

Specific: Task objective must be specific.
Measurable: Task objective must be measurable.
Aligned: Task objective must be achievable and aligned with overall project goal.
Realistic: Task objective must be realistic and relevant to the organization.
Timed: Task objective must have a time basis.

If a task has the above intrinsic characteristics, then the function of communicating the task will more likely lead to personnel cooperation.

Communication

Communication makes working together possible. The communication function of project management involves making all those concerned become aware of project requirements and progress. Those who will be affected by the project directly or indirectly, as direct participants or as beneficiaries, should be informed as appropriate regarding the following:

- Scope of the project;
- Personnel contribution required;
- Expected cost and merits of the project;
- Project organization and implementation plan;
- Potential adverse effects if the project should fail;
- Alternatives, if any, for achieving the project goal;
- Potential direct and indirect benefits of the project.

The communication channel must be kept open throughout the project life cycle. In addition to internal communication, appropriate external sources should also be consulted. The project manager must demonstrate the following:

- Exude commitment to the project.
- Utilize the communication responsibility matrix.
- Facilitate multi-channel communication interfaces.
- Identify internal and external communication needs.
- Resolve organizational and communication hierarchies.
- Encourage both formal and informal communication links.

When clear communication is maintained between management and employees and among peers, many project problems can be averted. Project communication may be carried out in one or more of the following formats:

- One-to-many
- One-to-one
- Many-to-one
- Written and formal
- Written and informal
- Oral and formal
- Oral and informal
- Nonverbal gestures.

Good communication is affected when what is implied is perceived as intended. Effective communications are vital to the success of any project. Despite the awareness that proper communications form the blueprint for project success, many organizations still fail in their communications functions. The study of communication is complex. Factors that influence the effectiveness of communication within a project organization structure include the following.

1. *Personal perception*. Each person perceives events on the basis of personal psychological, social, cultural, and experimental background. As a result, no two people can interpret a given event the same way. The nature of events is not always the critical aspect of a problem situation. Rather, the problem is often the different perceptions of the different people involved.
2. *Psychological profile*. The psychological makeup of each person determines personal reactions to events or words. Thus, individual needs and level of thinking will dictate how a message is interpreted.
3. *Social environment*. Communication problems sometimes arise because people have been conditioned by their prevailing social environment to interpret certain things in unique ways. Vocabulary, idioms, organizational status, social stereotypes, and economic situation are among the social factors that can thwart effective communication.
4. *Cultural background*. Cultural differences are among the most pervasive barriers to project communications, especially in today's multinational

organizations. Language and cultural idiosyncrasies often determine how communication is approached and interpreted.

5. *Semantic and syntactic factors.* Semantic and syntactic barriers to communications usually occur in written documents. Semantic factors are those that relate to the intrinsic knowledge of the subject of the communication. Syntactic factors are those that relate to the form in which the communication is presented. The problems created by these factors become acute in situations where response, feedback, or reaction to the communication cannot be observed.

6. *Organizational structure.* Frequently, the organization structure in which a project is conducted has a direct influence on the flow of information and, consequently, on the effectiveness of communication. Organization hierarchy may determine how different personnel levels perceive a given communication.

7. *Communication media.* The method of transmitting a message may also affect the value ascribed to the message and consequently, how it is interpreted or used. The common barriers to project communications are

- Inattentiveness
- Lack of organization
- Outstanding grudges
- Preconceived notions
- Ambiguous presentation
- Emotions and sentiments
- Lack of communication feedback
- Sloppy and unprofessional presentation
- Lack of confidence in the communicator
- Lack of confidence by the communicator
- Low credibility of communicator
- Unnecessary technical jargon
- Too many people involved
- Untimely communication
- Arrogance or imposition
- Lack of focus.

Some suggestions on improving the effectiveness of communication are presented next. The recommendations may be implemented as appropriate for any of the forms of communications listed earlier. The recommendations are for both the communicator and the audience.

1. Never assume that the integrity of the information sent will be preserved as the information passes through several communication channels. Information is generally filtered, condensed, or expanded by the receivers before relaying it to the next destination. When preparing a communication that needs to pass through several organization structures, one safeguard is to compose the original information in a concise form to minimize the need for re-composition of the project structure.

2. Give the audience a central role in the discussion. A leading role can help make a person feel a part of the project effort and responsible for the projects' success. They can then have a more constructive view of project communication.

3. Do homework and think through the intended accomplishment of the communication. This helps eliminate trivial and inconsequential communication efforts.

4. Carefully plan the organization of the ideas embodied in the communication. Use indexing or points of reference whenever possible. Grouping ideas into related chunks of information can be particularly effective. Present the short messages first. Short messages help create focus, maintain interest, and prepare the mind for the longer messages to follow.

5. Highlight why the communication is of interest and how it is intended to be used. Full attention should be given to the content of the message with regard to the prevailing project situation.

6. Elicit the support of those around you by integrating their ideas into the communication. The more people feel they have contributed to the issue, the more expeditious they are in soliciting the cooperation of others. The effect of the multiplicative rule can quickly garner support for the communication purpose.

7. Be responsive to the feelings of others. It takes two to communicate. Anticipate and appreciate the reactions of members of the audience. Recognize their operational circumstances and present your message in a form they can relate to.

8. Accept constructive criticism. Nobody is infallible. Use criticism as a springboard to higher communication performance.

9. Exhibit interest in the issue in order to arouse the interest of your audience. Avoid delivering your messages as a matter of a routine organizational requirement.

10. Obtain and furnish feedback promptly. Clarify vague points with examples.

11. Communicate at the appropriate time, at the right place, to the right people.

12. Reinforce words with positive action. Never promise what cannot be delivered. Value your credibility.

13. Maintain eye contact in oral communication and read the facial expressions of your audience to obtain real-time feedback.

14. Concentrate on listening as much as speaking. Evaluate both the implicit and explicit meanings of statements.

15. Document communication transactions for future references.

16. Avoid asking questions that can be answered yes or no. Use relevant questions to focus the attention of the audience. Use questions that make people reflect upon their words, such as "How do you think this will work?" compared to "Do you this will work?"

17. Avoid patronizing the audience. Respect their judgment and knowledge.

18. Speak and write in a controlled tempo. Avoid emotionally charged voice inflections.

19. Create an atmosphere for formal and informal exchange of ideas.
20. Summarize the objectives of the communication and how they will be achieved.

Within the framework of Triple C, a communication responsibility matrix contains the linking of sources of communication and targets of communication. Cells within the matrix indicate the subject of the desired communication. There should be at least one filled cell in each row and each column of the matrix. This assures that each individual of a department has at least one communication source or target associated with them. With a communication responsibility matrix, a clear understanding of what needs to be communicated to whom can be developed. Communication in a project environment can take any of several forms. The specific needs of a project may dictate the most appropriate mode. Three popular computer communication modes are discussed next in the context of communicating data and information for project management.

Simplex communication. This is a unidirectional communication arrangement in which one project entity initiates communication to another entity or individual within the project environment. The entity addressed in the communication does not have mechanism or capability for responding to the communication. An extreme example of this is a one-way, top-down communication from top management to the project personnel. In this case, the personnel have no communication access or input to top management. A budget-related example is a case where top management allocates budget to a project without requesting and reviewing the actual needs of the project. Simplex communication is common in authoritarian organizations.

Half-duplex communication. This is a bi-directional communication arrangement whereby one project entity can communicate with another entity and receive a response within a certain time lag. Both entities can communicate with each other but not at the same time. An example of half-duplex communication is a project organization that permits communication with top management without a direct meeting. Each communicator must wait for a response from the target of the communication. Request and allocation without a budget meeting is another example of half-duplex data communication in project management.

Full-duplex communication. This involves a communication arrangement that permits a dialogue between the communicating entities. Both individuals and entities can communicate with each other at the same time or face-to-face. As long as there is no clash of words, this appears to be the most receptive communication mode. It allows participative project planning in which all project personnel have an opportunity to contribute to the planning process.

Each member of a project team needs to recognize the nature of the prevailing communication mode in the project. Management must evaluate the prevailing communication structure and attempt to modify it if necessary to enhance project functions. An evaluation of who is to communicate with whom about what may help improve the project data/information communication process. A communication matrix may include notations about the desired modes of communication between individuals and groups in the project environment.

Cooperation

The cooperation of the project personnel must be explicitly elicited. Merely voicing consent for a project is not enough assurance of full cooperation. The participants and beneficiaries of the project must be convinced of the merits of the project. Some of the factors that influence cooperation in a project environment include personnel requirements, resource requirements, budget limitations, past experiences, conflicting priorities, and lack of uniform organizational support. A structured approach to seeking cooperation should clarify the following:

- Cooperative efforts required;
- Precedents for future projects;
- Implication of lack of cooperation;
- Criticality of cooperation to project success;
- Organizational impact of cooperation;
- Time frame involved in the project;
- Rewards of good cooperation.

Cooperation is a basic virtue of human interaction. More projects fail due to a lack of cooperation and commitment than any other project factors. To secure and retain the cooperation of project participants, you must elicit a positive first reaction to the project. The most positive aspects of a project should be the first items of project communication. For project management, there are different types of cooperation that should be understood.

Functional cooperation. This is cooperation induced by the nature of the functional relationship between two groups. The two groups may be required to perform related functions that can only be accomplished through mutual cooperation.

Social cooperation. This is the type of cooperation effected by the social relationship between two groups. The prevailing social relationship motivates cooperation that may be useful in getting project work done.

Legal cooperation. Legal cooperation is the type of cooperation that is imposed through some authoritative requirement. In this case, the participants may have no choice other than to cooperate.

Administrative cooperation. This is cooperation brought on by administrative requirements that make it imperative that two groups work together on a common goal.

Associative cooperation. This type of cooperation may also be referred to as collegiality. The level of cooperation is determined by the association that exists between two groups.

Proximity cooperation. Cooperation due to the fact that two groups are geographically close is referred to as proximity cooperation. Being close makes it imperative that the two groups work together.

Dependency cooperation. This is cooperation caused by the fact that one group depends on another group for some important aspect. Such dependency is usually of a mutual, two-way nature. One group depends on the other for one thing, while the latter group depends on the former for some other thing.

Imposed cooperation. In this type of cooperation, external agents must be employed to induced cooperation between two groups. This is applicable for cases where the two groups have no natural reason to cooperate. This is where the approaches presented earlier for seeking cooperation can became very useful.

Lateral cooperation. Lateral cooperation involves cooperation with peers and immediate associates. Lateral cooperation is often easy to achieve because existing lateral relationships create an environment that is conducive for project cooperation.

Vertical cooperation. Vertical or hierarchical cooperation refers to cooperation that is implied by the hierarchical structure of the project. For example, subordinates are expected to cooperate with their vertical superiors.

Whichever type of cooperation is available in a project environment, the cooperative forces should be channeled toward achieving project goals. Documentation of the prevailing level of cooperation is useful for winning further support for a project. Clarification of project priorities will facilitate personnel cooperation. Relative priorities of multiple projects should be specified so that a priority to all groups within the organization. Some guidelines for securing cooperation for most projects are

- Establish achievable goals for the project.
- Clearly outline the individual commitments required.
- Integrate project priorities with existing priorities.
- Eliminate the fear of job loss due to industrialization.
- Anticipate and eliminate potential sources of conflict.
- Use an open-door policy to address project grievances.
- Remove skepticism by documenting the merits of the project.

Commitment. Cooperation must be supported with commitment. To cooperate is to support the ideas of a project. To commit is to willingly and actively participate in project efforts again and again through the thick and thin of the project. Provision of resources is one way that management can express commitment to a project.

Coordination

After the communication and cooperation functions have successfully been initiated, the efforts of the project personnel must be coordinated. Coordination facilitates harmonious organization of project efforts. The construction of a responsibility chart can be very helpful at this stage. A responsibility chart is a matrix consisting of columns of individual or functional departments and rows of required actions. Cells within the matrix are filled with relationship codes that indicate who is responsible for what. Table 10.1 illustrates an example of a responsibility matrix for the planning for a seminar program. The matrix helps avoid neglecting crucial communication requirements and obligations. It can help resolve questions such as the following:

- Who is to do what?
- How long will it take?
- Who is to inform whom of what?
- Whose approval is needed for what?
- Who is responsible for which results?
- What personnel interfaces are required?
- What support is needed from whom and when?

TABLE 10.1

Example of Responsibility Matrix for Project Coordination

TASKS	Person Responsible				Status of Task			
	Staff A	Staff B	Staff C	Mgr	31-Jan	15-Feb	28-Mar	21-Apr
Brainstorming Meeting	R	R	R	R	D			
Identify Speakers			R			O		
Select Seminar Location	I	R	R			O		
Select Banquet Location	R	R				D		
Prepare Publicity Materials		C	R	I	O	O	D	
Draft Brochures		C	R					D
Develop Schedule		R				L	L	
Arrange for Visual Aids		R			L	L	L	
Coordinate Activities		R					L	
Periodic Review of Tasks	R	R	R	S				D
Monitor Progress of Program	C	R	R			O	L	
Review Program Progress	R				O	O	L	L
Closing Arrangements	R						L	
Post-Program Review and Evaluation	R	R	R	R			D	

Responsibility Codes: **Task Codes:**

R = Responsible   D = Done
I = Inform   O = On Track
S = Support   L = Late
C = Consult

Conflict Resolution Using the Triple C Approach

Conflicts can and do develop in any work environment. Conflicts, whether intended or inadvertent, prevent an organization from getting the most out of the workforce. When implemented as an integrated process, the Triple C model can help avoid conflicts in a project. When conflicts do develop, it can help in resolving the conflicts. The key to conflict resolution is open and direct communication, mutual cooperation, and sustainable coordination. Several sources of conflicts can exist in a project. Some of these are discussed below.

Schedule conflict. Conflicts can develop because of improper timing or sequencing of project tasks. This is particularly common in large multiple projects. Procrastination can lead to having too much to do at once, thereby creating a clash of project functions and discord among project team members. Inaccurate estimates of time requirements may lead to infeasible activity schedules. Project coordination can help avoid schedule conflicts.

Cost conflict. Project cost may not be generally acceptable to the clients of a project. This will lead to project conflict. Even if the initial cost of the project is acceptable, a lack of cost control during implementation can lead to conflicts. Poor budget allocation approaches and the lack of a financial feasibility study will cause cost conflicts later on in a project. Communication and coordination can help prevent most of the adverse effects of cost conflicts.

Performance conflict. If clear performance requirements are not established, performance conflicts will develop. Lack of clearly defined performance standards can lead each person to evaluate their own performance based on personal value judgments. In order to uniformly evaluate quality of work and monitor project progress, performance standards should be established by using the Triple C approach.

Management conflict. There must be a two-way alliance between management and the project team. The views of management should be understood by the team. The views of the team should be appreciated by management. If this does not happen, management conflicts will develop. A lack of a two-way interaction can lead to strikes and industrial actions, which can be detrimental to project objectives. The Triple C approach can help create a conducive dialogue environment between management and the project team.

Technical conflict. If the technical basis of a project is not sound, technical conflict will develop. New industrial projects are particularly prone to technical conflicts because of their significant dependence on technology. Lack of a comprehensive technical feasibility study will lead to technical conflicts. Performance requirements and systems specifications can be integrated through the Triple C approach to avoid technical conflicts.

Priority conflict. Priority conflicts can develop if project objectives are not defined properly and applied uniformly across a project. Lack of a direct project definition can lead each project member to define their own goals which may be in conflict with the intended goal of a project. Lack of consistency of the project mission is another potential source of priority conflicts. Over-assignment of responsibilities with no guidelines for relative significance levels can also lead to priority conflicts. Communication can help defuse priority conflict.

Resource conflict. Resource allocation problems are a major source of conflict in project management. Competition for resources, including personnel, tools, hardware, software, and so on, can lead to disruptive clashes among project members. The Triple C approach can help secure resource cooperation.

Power conflict. Project politics lead to a power play which can adversely affect the progress of a project. Project authority and project power should be clearly delineated. Project authority is the control that a person has by virtue of their functional post. Project power relates to the clout and influence which a person can exercise due to connections within the administrative structure. People with popular personalities can often wield a lot of project power in spite of low or nonexistent project authority. The Triple C model can facilitate a positive marriage of project authority and power to the benefit of project goals. This will help define clear leadership for a project.

Personality conflict. Personality conflict is a common problem in projects involving a large group of people. The larger the project, the larger the size of the management team needed to keep things running. Unfortunately, a larger management team creates a greater opportunity for personality conflicts. Communication and cooperation can help defuse personality conflicts. In summary, conflict resolution through Triple C can be achieved by observing the following guidelines:

1. Confront the conflict and identify the underlying causes.
2. Be cooperative and receptive to negotiation as a mechanism for resolving conflicts.
3. Distinguish between proactive, inactive, and reactive behaviors in a conflict situation.
4. Use communication to defuse internal strife and competition.
5. Recognize that short-term compromise can lead to long-term gains.
6. Use coordination to work toward a unified goal.
7. Use communication and cooperation to turn a competitor into a collaborator.

It is the little and often neglected aspects of a project that lead to project failures. Several factors may constrain the project implementation. All the relevant factors can be evaluated under the Triple C model right from the project-initiation stage.

Application of Triple C to Environmental Projects

Having now understood the intrinsic elements of Triple C, we can see how and where it could be applicable to environmental project management. Communication explains project scope and requirements through the stages of planning, organizing, scheduling, and control. Cooperation is required to get human resource buy-in and stakeholder endorsement across all facets of planning, organizing, scheduling, and control. Coordination facilitates adaptive interfaces over all the elements of planning, organizing, scheduling, and control. The Triple C model should be implemented as an iterative loop process that moves a project through the communication, cooperation, and coordination functions.

DMAIC and Triple C

Many organizations now explore Six Sigma DMAIC (Define, Measure, Analyze, Improve, and Control) methodology and associated tools to achieve better project performance. Six Sigma means six standard deviations from a statistical performance average. The Six Sigma approach allows for no more than 3.4 defects per million parts in manufactured goods or 3.4 mistakes per million activities in a service operation. To explain the effect of the Six Sigma approach, consider a process that is 99% perfect. That process will produce 10,000 defects per million parts. With six sigma, the process will need to be 99.99966% perfect in order to produce only 3.4 defects per million. Thus, Six Sigma is an approach that moves a process toward perfection. Six Sigma, in effect, reduces variability among products produced by the same process. By contrast, Lean approach is designed to reduce/eliminate waste in the production process.

Six Sigma provides a roadmap for the five major steps of DMAIC, which are applicable to the planning and control steps of project management. We cannot improve what we cannot measure. Triple C provides a sustainable approach to obtaining cooperation and coordination for DMAIC during improvement efforts. DMAIC requires project documentation and reporting, which coincide with project control requirements.

This chapter has presented a general introduction to the Triple C approach, focusing on communication, cooperation, and coordination. A summary of lessons to be inferred from a Triple C approach can be described as follows:

- Use proactive planning to initiate project functions.
- Use preemptive planning to avoid project pitfalls.
- Use meetings strategically. Meeting is not *work*. Meeting should be done to facilitate work.
- Use project assessment to properly frame the problem, adequately define the requirements, continually ask the right questions, cautiously analyze risks, and effectively scope the project.
- Be bold to terminate a project when termination is the right course of action. Every project needs an exit plan. In some cases, there is victory in capitulation.

The applicability and sustainability of the Triple C approach is summarized below:

1. For effective communication, create good communication channels.
2. For enduring cooperation, establish partnership arrangements.
3. For steady coordination, use a workable organization structure.

Further details on Triple C can be found in Badiru (2008).

REFERENCES

Badiru, A. B. (1987, May). *Communication, cooperation, coordination: The triple C of project management.* Proceedings of 1987 IIE Spring Conference, Washington, DC, pp. 401–404.

Badiru, A. B. (2008). *Triple C model of project management: Communication, cooperation, and coordination.* Taylor & Francis/CRC Press.

Badiru, A. B., Foote, B. L., Leemis, L., Ravindran, A. R., & Williams, L. (1993, February). Recovering from a crisis at tinker air force base. *PM Network, 7*(2), 10–23.

Index

Printed in the United States
by Baker & Taylor Publisher Services